What people are saying about

Escaping from Eden

Escaping from Eden takes us on a journey that we will never forget. The incredible possibilities of our own existence. Thought provoking

George Noory, Host of the national radio show *Coast to Coast AM*

Paul is doing a courageous service...to give us a new perspective on the creation and engineering of man.

Sean Stone, Filmmaker and Media Host

Escaping from Eden

Does Genesis teach that the human race was created by God or engineered by ETs?

Escaping from Eden

Does Genesis teach that the human race was created by God or engineered by ETs?

Paul Wallis

AXIS MUNDI
BOOKS

Winchester, UK
Washington, USA

First published by Axis Mundi Books, 2020
Axis Mundi Books is an imprint of John Hunt Publishing Ltd., 3 East Street, Alresford, Hampshire SO24 9EE, UK
office@jhpbooks.com
www.johnhuntpublishing.com
www.zero-books.net

For distributor details and how to order please visit the 'Ordering' section on our website.

Note
All Bible quotations are a free translation except where otherwise stated.

ISBN: 978 1 78904 387 7
978 1 78904 388 4 (ebook)
Library of Congress Control Number: 2019937316

A CIP catalogue record for this book is available from the British Library.

Design: Stuart Davies

Printed and bound in Great Britain by
TJ Books Limited, Padstow, Cornwall

I dedicate this book with love to my beautiful family,
Ruth, Evie, Ben and Caleb – and to all seekers of truth
and consciousness.

Acknowledgements

This book would not be in your hands without the help and encouragement of an army of people. My thanks especially to my family for their great patience in supporting the writing of this book and to my literary colleagues Michael Mann, Andrew James Wells, Krystina Kellingley, Dominic C James, Brian Mountford, Mary Flatt, Stuart Davies, Beccy Conway, Nick Welch, Maria Barry, and John Hunt – also to George Noory, Regina Meredith, Sean Stone and Erich Von Daniken whose work and encouragement helped midwife this book into being. Thank you!

INTRODUCTION & CHAPTER ONE

INJURIES & ANOMALIES

Another flash.

"Nick Hallatt of the BBC. Can you comment on Pope Francis' response to your book and were you surprised?"

I willed my glasses not to fog up in the face of flashing cameras and rapid-fire questions. I did not want to see my face in some newspaper column the next morning looking like a startled rabbit.

"Thank you, Nick. Well, I know that since Pope Benedict, and now with Pope Francis the Vatican has been widening its theological doors on this topic. So I really welcome that. I think that's a real encouragement for all people to explore the kind of topics my book touches on. But I have to say, at this point I haven't yet had the opportunity to really reflect on the papal statement. It was only released half an hour ago – just as I was on my way to you!"

Flash.

"Michelle Block NPR – Mr Wallis, you're making huge claims about Bible translation, archaeology, DNA research. According to my research you don't have qualifications in any of these fields, so why would a reader need to take the claims of your book seriously."

"Thank you, Michelle. That's a great question! I guess a core part of what people look for from their pastors and preachers is for us to try and make sense of the Scriptures that we preach from. A big part of that is carrying people's questions, wrestling with them and doing that openly. I've been doing that for more than thirty years. And that's really what I'm doing in this book.

"Now there are plenty of times when I have to stand on others' shoulders. That's why in my book I go to some world-class academics and researchers, people who have given their lives to their fields. So, a lot of my questions have taken me to those people."

You can think of my book as a kind of documentary, sharing the preacher's journey with the reader. My hope is that this little book will be a way in to some of these amazing areas of study – especially for people of faith."

Another flash. Just as I was looking down at my notes.

"Ted Avery, Fox News – Paul, you say it's for people of faith but in what way can you call yourself a Christian when in fact your book pulls the rug out from under 2000 years of Biblical interpretation? The two theologians you hold up were both condemned as heretics. Doesn't that mean your position is built on heresy? Can you name a single Christian leader today who is willing to raise their hand and support any one of your book's ridiculous conclusions – other than Pope Francis?"

I have to say, this line of questioning was not a complete surprise to me. Faith and new ideas do not always make easy company. I thought of my friend, Vince. He was a senior theologian for many years for a heterodox Christian sect. He found himself out on his ear when his careful study of the Bible led him to conclude that their particular sect had not got all its Biblical translation right. Better translation meant that their sect could not claim to have exclusive access to the kingdom of God.

When Vince sounded out his senior colleagues, they all told him the same thing:

"We know! We realized that years ago. Just don't mention what you're thinking to headquarters. Without that brick in the wall they feel the whole house will come falling down. Don't do it Vince or they'll kick you out. You'll be shunned. We couldn't do it. All our friends and family are in this movement. It's our life."

Mainstream denominations aren't so different either. A hundred years ago a Baptist minister, having served more than a decade as a professor of Semitic languages, suddenly found himself disendorsed and out of a job. That was for publishing papers that only went half as far as my new book.

I don't say that in any way to be judgey about church and faith communities, because at the end of the day we are all creatures of habit. It's not easy to change our minds. Not one of us is ever really prepared to wake up in a universe that's different to the one we fell asleep in. I knew that my book would ruffle feathers and maybe even lose me a friend or two. My editor had run theological gauntlets like this himself before. So I was pleased to have him at my side as the questions continued.

Flash.

"Erm. Hugh Grant from Horse and Hound magazine..."

It was all too imaginable given the topic of my new book. Perhaps I should have written another gentle devotional book like my previous offerings on Celtic spirituality and Eastern Orthodox mysticism. They were nice books and they didn't cost me any friends! On the other hand, I just couldn't not-publish this book, even if it meant running a gauntlet like the one I had just imagined. I had to publish it because, firstly, that's what addictive writers do. They share the journey. And secondly, because I've learned, just as Neo does in *The Matrix,* that once you've taken the red pill there's no going back.

* * *

INJURIES & ANOMALIES

I had seen these anomalies before. I knew that something was off and I could see how these glitches might just throw spanners into the works of everything I was trained in. But never before had my whole working future ridden on what they meant. Somehow, my pace of work always kept me just that bit too busy to give the glitches much attention. This time around it was different.

The upside of my Ultimate Frisbee match with the youth group was that I was on the winning team – and we won by a country mile. The downside was having to spend the next I-don't-know-how-many weeks with my lower right leg bolted into a *"portable"* traction device. Just a few weeks before, my wife Ruth and I had invested in a shipping crate cabin to adorn our driveway. The idea was to provide some *"tiny house"* accommodation for our guests and AirBNB it for a bit of extra cash. Right now, I was appreciating it for myself as a place of quiet and seclusion to help me recover and to get healed up.

Before the Ultimate Frisbee incident, I had been preaching through the book of Genesis – one of my favorite books. Now that the universe had gifted me with a period of quiet, I could study it afresh without the background pressure of having only six days to create the next sermon, with sense made of every detail and every loose end tied. Churches generally like their pastors to make sense of the Scriptures they preach, and my congregation at the Church on the Range in Victoria, Australia was no exception. As much as anywhere the good people of this beautiful part of the world appreciate being challenged and stretched to a degree, but as in most churches there is a familiar canon of stories which our people expect to hear reaffirmed on a regular basis. There is comfort in the familiar rhythms of the old, old story. It's a story which speaks of an Almighty God who creates light, space, energy, matter, stars and planets. Out of nothing he forms land and sea, vegetation and animals, and ultimately people like you and me. And the plot rolls on from there. Except for the anomalies. Those little red flags, all signaling that something is awry.

With my time moving more slowly, I became aware that every verse that didn't fit, every word that didn't make sense, was somehow harder to brush off. Each time I sat down to read through the book of Genesis, the same anomalous verses kept getting in my face and flagging me down, as if to say, *"Paul! Stop! Don't read any further. You've got the story wrong!"*

According to the history of scientific discovery, anomalies are supposed to be our friends. They're the little clues that our metanarrative is off. They beckon us back to the data to take another look. When you're overscheduled and don't have the time for them, you tend to see anomalies in your data as annoyances and want to quickly dismiss them or explain them away. It's the same with the Bible. Take a good long look and the many anomalous verses of Scripture begin to reveal themselves as something altogether more enigmatic. Give them enough attention and you'll realize they are the portals to another world.

If you have ever read the first eleven chapters of the book of Genesis in the Bible – the stories of beginnings – then you probably have an idea of what the anomalies are that were derailing the preparations for my next sermon series.

In this chapter I'm going to invite you to read over my shoulder the random notes I made as these portals in the Scriptures began to open up. But I'll warn you, we'll be diving into the deep end.

The anomalies appear early on. In fact, our familiar story gets messed up within the very first verse of the Bible.

Genesis 1:1 In the beginning God created? (NIV)

When I read this verse in English there's no problem. With my interlinear Bible open, the Hebrew text on one side and the Greek of the Septuagint on the other, I can't escape a rather big question:

Why is this word, elohim, which is translated as God, shaped like a plural noun? How come it's a plural if there's only one God?

Genesis 1:26 Let us make? (NIV)

Wait a minute! Who is this "us"? This is the dawn of time, before any intelligent creature has been named. So who or what are the others that God is talking to?

Genesis 1:26 Let us make human beings to look like us – in the image and likeness of ourselves?

The footnotes in my New Jerusalem Bible say, "Image is a concrete term, implying a physical resemblance – like that between Adam and his son." So these elohim are plural, creative and physical? What's that about?

Genesis 2:10–14 A river watering the garden flowed from Eden; from there it was separated into four headwaters. The name of the first is the Pishon; it winds through the entire land of Havilah, where there is gold. The gold of that land is good. Aromatic resin and onyx are also there. The name of the second river is the Gihon; it winds through the entire land of Cush. The name of the third river is the Tigris; it runs along the east side of Ashur. And the fourth river is the Euphrates. (NIV)

If, somewhere in Eden, the humans have been provided with a garden to meet all their needs, why is the Bible telling me the geographical location of key mineral deposits – Havilah for gold, resin and onyx – and the gold is high-grade. Who needs those? How is that relevant?

Genesis 2:17...but you must not eat from the tree of the knowledge of good and evil, for when you eat from it you will certainly die. (NIV)

How is the knowledge of good and evil a bad thing for humans to have? If God wanted humans to be capable of free choice and love, then surely this kind of moral awareness is absolutely essential. If the humans are incapable of distinguishing good from bad then how can they choose the good, or be held culpable if they choose the bad?

Surely, if the man and woman have no moral awareness then God has set them up to fail. They can only fail. And if they fail how can he hold them responsible? That doesn't make sense. On top of that, how can death be a just punishment from a loving God? It doesn't fit the "crime." Something's wrong with that picture.

Genesis 2:21 So the man gave names to all the livestock, the birds in the sky and all the wild animals. But for Adam no suitable helper was found. So the Lord God caused the man to fall into a deep sleep; and while he was sleeping, he took one of the man's ribs and then closed up the place with flesh. (NIV)

Isn't this a little short-sighted? If the Almighty is the source of all wisdom, how can a female of the species be an afterthought? How can woman be given as the man's helper only after a line of pets has proven inadequate?

Genesis 3:1 Now the Snake...(NIV)

OK, who is this? Clearly, he's not a snake when he turns up. He is intelligent. He has arms and legs and can speak human. He's a significant player but he arrives on the scene without any kind of introduction or explanation! So what is he? He is not the Almighty and he is not human. Has any other kind of being been named so far? Did I miss something?

Did I skip the verse that explains this other kind of being?

I wouldn't be the first person to miss the glaringly obvious. Come to think about it, almost every time Jesus quotes the Hebrew Scriptures, he says, *"Friends, look again! I think you may have missed something!"*

(The fact is I had missed something. Yet I was soon to find it, hidden in plain sight, in the very first verses of the book of Genesis.)

Genesis 3:16 In pain you will give birth to children...(NIV)

Is childbearing a punishment or a consequence? No childbirth has occurred prior to this moment. In the flow of the story, painful childbirth is the only kind – and it comes after the humans gain conscience, self-awareness and sexuality – naturally!

Genesis 3:22 And Yahweh Elohim said, *"See the human has become like one of us!"* (NIV)

Just a second! One of who? Who is God "like" that we humans would be like "them?" This isn't a literary tick. It's not a royal "we" or an author's "we." This really is a plural!

Genesis 4:14 "I will be a restless wanderer on the earth and whoever finds me will kill me!" (NIV)

Excuse me! Who are these other people, these "whoevers" outside the garden who are going to kill Cain if they meet him? They can't be children of Adam and Eve. So who?

Genesis 6–9 The animals went in two by two. Right?

Apparently not! I can see that Genesis 1 and 2 put two creation narratives side by side. Genesis 6 to 9 looks like it has taken two flood accounts and woven them together. In the one version the animals enter in pairs. In the other they've worked out a more complicated system.

Just a reminder that Genesis has taken from a number of sources and interwoven them to create the familiar version. So some texts tell stories of the elohim. Others speak of "Yahweh" – the Holy Name of God, revealed to Moses in a later age. Scholars call this second narrator "J" (or more traditionally, "Moses").

I remember reading up at college on eighteenth century Bible scholars like Jean Astruc, Karl Heinrich Graf and Julius Wellhausen, who talked about the redactors (scissors-and-paste editors) who put Genesis into its current shape. These academics might not have entirely nailed it but it's easy to see how J inserts the Holy Name into Genesis.

Sometimes J replaces "elohim" with "Yahweh" (e.g. Genesis 11:6–7 and 18:21–19:1). Other times he just adds the name Yahweh to elohim in an elohim story (e.g. Genesis 3:22).

I can see why he would do that. J is telling the reader to see the hand of Yahweh in the drama of the stories.

How did I miss the implication of this? Because the very presence of the post-Moses name of Yahweh in Genesis tells me that we are not reading the original version of Genesis!

J isn't hiding what he is doing. When he adds the Holy Name, J shows to the reader that he is taking an even older story (which was probably a known story, a written or oral tradition), and in plain sight he alters it!

So now I have to ask, "What about the original version of these stories? What texts was J working with? And if they weren't Yahweh stories, and if the elohim in them were plural, were the original versions of the Genesis stories God-stories at all?"

Genesis 6:1–4 The Nephilim were on the earth in those days—and also afterward—when the sons of God went to the daughters of humans and had children by them. They were the heroes of old, men of renown. (NIV)

Wait a minute! If the flood destroyed all living creatures on the land, where did the Nephilim (giants) go that they could reappear after the flood?

Genesis 6:1–4 When human beings began to increase in number on the earth and daughters were born to them, the sons of God *(benei elohim)* **saw that the daughters of humans were beautiful, and they married any of them they chose.** (NIV)

Who were the benei elohim (often translated as sons of God)? And what is different about them that their relations with human females produces giants?

I have read somewhere that some commentators reckon *benei elohim is talking about local powers or kings or elites. If* land-barons were taking girls by force that's certainly a problem. But that explanation makes no sense of their noticing the "daughters of men" – *as if human girls were some kind of novelty. Neither would their relations produce giants. And would that really be cause for a catastrophic, genociding flood? It doesn't quite hang together. Perhaps there's more to these prehistoric "land-barons" than meets the eye! Maybe these prehistoric communities were being ruled over by something else?? We're in the Twilight Zone here, because I don't think these benei elohim are human!*

I know that some interpret benei elohim as angels. But doesn't Hebrews in the New Testament rule that out? It says:

For to which of the angels did God ever say, "You are my Son; today I have become your Father"? Or again, "I will be his Father, and he will be my Son"?

Maybe benei elohim is a more figurative kind of expression. Could these entities be referred to as "sons of elohim" in the same way we refer to "sons and daughters of the Mayflower" or "sons and daughters of the revolution?" In other words, perhaps it's not literal. Perhaps it's an idiom. That way "benei elohim" could mean "the God-like ones" or "elohim-kind." Just as we have "humans" existing within "humankind" maybe "elohim" exist within "benei elohim."

Whatever the benei elohim are exactly, they're not God, and they're not human. What they are has not been explained. For some strange reason, J assumes we all know about them. But from where! What other sources does J think we've been reading?

Genesis 20:13 "The Elohim caused me to wander from my father's house."

If elohim is supposed to be a general name for a god or a proper name for the one True God (Yahweh) then why does the word get given plural attributives and verb forms? In this text Abraham uses a plural elohim with a plural verb.

I have read commentators who say that Abraham wants to sound like a polytheist because he is talking to a polytheist. Really? And the editor didn't iron that out?

I'm finding these plural issues even further into Genesis.

In **Genesis 35:7** *it says, "And there [Jacob] built an altar and called the place Bethel, because there elohim appeared (plural verb) to him when he was fleeing his brother."*

Same in **Genesis 11:7** Yahweh says *"Come let us go down and let us confuse (plural verb) their language there..."*

Genesis 22 *Abraham's near sacrifice of his son Isaac has never made sense to me. Why would a loving God, the God who had promised Abraham a family line through his son, Isaac, then ask him to mimic the absolute worst of pagan religion and sacrifice his first-born son? A plural non-divine elohim would make a lot more sense than J's edit of this episode. In this passage elohim and Yahweh actually play on opposite teams! It begins when elohim tell Abraham to sacrifice his first-born and only son...*

"It happened some time later that elohim put Abraham to the test...Elohim said, "Take your son, your only son...offer him as a burnt offering...Abraham stretched out his hand and took the knife to kill his son."

Isn't this how other "gods" proved they had exacted total, unquestioning obedience from their human subjects? The Greek islands, Phoenicia, the near East, Mesoamerica were all areas where child sacrifice was expected by the local elohim. What must those elohim have been to have produced such inhuman behavior among their subjects? When elohim demand it of Abraham, the stakes are high. Isaac is the son from whom the people of Israel are supposed to descend. Hence Yahweh acts to prevent the killing:

"From out of heaven the messenger of Yahweh shouted and called out to Abraham, saying, 'Abraham, Abraham! Do not raise your hand against the boy!'"

If it was already well-known, this story would pose a problem for J. There are too many gods in it! To turn it into a monotheistic text J has translated elohim to mean the same as Yahweh. He equates them by inserting the last word of the story's last sentence. Yahweh says:

"Now I know that you fear elohim and did not withhold your son from...me."

With this neat addendum J turns it into a monotheistic text. But by equating Yahweh with the elohim who gave the filicidal order, J changes the story from one of gracious divine rescue to a cruel divine test. And it makes the story a moral nonsense. How am I supposed to feel about a deity who commands unquestioning obedience, even to the point of being willing to kill your own children? J has turned it into a story with only one "god" in it. But in so doing the "god" he portrays becomes a monster.

All this makes me think about what happened to F.D.Maurice back in the day. Maurice was Chaplain of King's College London in the 1850s. Then he published some papers questioning the morality of the traditional church doctrine of hell. He argued that eternal, conscious torment couldn't be the correct interpretation of Jesus' teaching because eternal punishment is so absurdly out of proportion with any human crime. An infinite period of torture for a finite crime is nothing but cruel and unjust.

Maurice was sacked and accused of heresy. Strange because he didn't actually contradict anything in the official creeds of the Church. The University said it was worried that Maurice would "unsettle the theological students" by putting a moral argument as a case for a theological position. What that really means is that Maurice broke three unspoken laws: "Thou shallt not question the Almighty. Thou shallt not expect thy God to adhere to any morality a mere mortal could understand. Thou shallt not have compassion on people thy God wishes to torment or kill."

They were more or less saying that if God can expect adherence to those three laws from Abraham then he can certainly expect them of a King's College Chaplain – or you or me! But if we don't question, expect, and have compassion, then we fail to notice if our god has become a monster.

If you want people to never question authority, especially those speaking for God, then J's version of the near sacrifice of Isaac does the job. But haven't we just had Royal Commissions in the UK and Australia showing the horrible abuses that result from that kind of theology. If you want to forbid any moral evaluation of supposedly divine actions, then J's account suffices. On the other hand, if we're looking for an undistorted image of God then the current version of Genesis becomes something we have to avoid or explain away.

Now that you've eavesdropped on my notes, you'll understand the wrestle I found myself wrestling. Everything hinged on this enigmatic and anomalous word – *elohim.*

As I continued to walk through the pages of Genesis, I could see that every time *elohim* and Yahweh are equated there is a kind of moral distortion that takes place and the devout reader is forced to excuse ways that appear lower than ours – not higher. I began to wonder if the apparent morality of God would lift consistently if we were to re-pluralize our understanding of *elohim*?

These were exactly the kinds of questions my library of commentaries liked to avoid, noting only briefly, explaining away unconvincingly, and moving on swiftly to easier points. So I decided to look to the historic Jewish commentators for a longer view. I reckoned these guys must have a couple of thousand years' advantage on the Christian commentators and presumably have a few more conclusions up their sleeves.

I soon found Joshua Ben Levi, writing in the third century AD and addressing the mystery of the Biblical conversations among the *elohim*. He said, *"With whom did God take counsel? With the works of heaven, he took counsel."*

A generation later Rabbi Ammi concluded, *"He took counsel with his own heart."* (Midrash Rabbah, Genesis 8,3,1 – tr.Jacob Neusner.)

OK, but that doesn't explain *"in the image and likeness of us"* or *"humans have become like one of us."*

The Aramaean commentary of the *Targum of Palestine* otherwise known as the *"Targum of Jonathan ben Uzziel"* proposes that the appearance of plural *elohim* doing the work of creation implies God acting in cooperation with a heavenly council. This council comprises himself and a number of powerful angels, whom God must have created, without mention, on the second day of creation. In Genesis 11 when *elohim* says, *"Let us go down,"* the commentator proposes that God must have been addressing *"the 70 angels who stand before him."*

OK but there is no mention of the creation of this body of angels in the text of Genesis. Wouldn't that be an odd omission?

Though it appeared to leave significant questions hanging, the Targum of Palestine pointed me to a phenomenon in the Bible that, somehow, I had never really noticed before. I must have read about it and just dismissed it as a creative flourish rather than anything real. I am talking about the heavenly council. Once you look, the heavenly council evidences itself in many pages of Scripture.

For instance:

"El takes his stand in the divine assembly, surrounded by the elohim he gives judgment." (Psalm 82:1)

"Who in the skies can compare with Yahweh? Who among the benei elohim can rival him? God, awesome in the great assembly of benei elohim, dreaded among all who surround him? Yahweh God of armies, who is like you?" (Psalm 89:6-8a)

"I saw Yahweh seated on his throne with the whole array of heaven standing by him, on his right and on his left. Yahweh said, 'Who will entice Ahab into marching to his death at Ramoth in Gilead?' at which some answered one way and some answered another. A spirit then came forward and stood before Yahweh and said, 'I will entice him...I will go and be a deceiving spirit in the mouths of all [Ahab's] prophets.'" (I Kings 22:19-23)

"One day when the benei elohim came to attend on Yahweh, among

them came the accuser." (Job 1:6)

The word *accuser* in the Job passage is often translated as *"Satan."* However, the word appears here not as a name but as a descriptor-noun with a definite article *ie "the accuser."* Nevertheless, the presence of an accuser among the *elohim,* apparently as a member of the heavenly council, is something that must raise an eyebrow or two! I was going to have to return to that later!

Joining these dots really had me scratching my head. How could I have read so much and yet missed so much? This was new to me. Here in the Bible is a heavenly council, populated by a diversity of entities, not all good, at least some of whom are spirits, and who are all *elohim-kind.* That's an interesting council!

After a time surveying the various behaviors of the word *elohim,* I began to wonder if in the Bible we are seeing not just a word with a diversity of usages, but a word in transition.

What I mean is, words change their meanings over time. A classic example from monastic history is the word *"chapter."* Originally the *chapter* referred to the portion of the monastic constitution which would be read once a week at a special gathering of members in a special assembly room. After a while *chapter* came to refer to the assembly room in which the constitution was read. After another while *chapter* came to refer to the group of people assembled in the room to hear the constitution read. Still later *chapter* came to mean the local gathering of any association of people.

So today if you read the word *chapter,* to know what it means you will need to let the date and context tell you which incarnation of *chapter* you're reading about! And to know why a group of people is called a *chapter* – you will need to know what it meant way back in the beginning.

Could it be that in the centuries spanned by the books of the Bible, we are watching the word *elohim* make a similar journey?

Perhaps the obvious is staring us in the face. The word is a plural form. So then, what if, in the beginning, before it became a proper name, *elohim* was simply a plural word?

If we read it that way, then without modifying the text one jot or tittle we could say that in Genesis 1 *elohim* doesn't *have* a heavenly council, *elohim is* the heavenly council.

If we read *elohim* as a simple plural – then the use of the word for other gods is no problem at all. The verses which accompany *elohim* with singular verb forms would then indicate moments where *elohim* are being referenced as a group or a collective.

For some real fire power in the department of Biblical translation, I turned to Dom Henry Wansbrough. Dom Henry was the supervising editor of the New Jerusalem Bible – one of the best Bible translations for the way it handles the names of God, and its superb scholarship and thorough footnotes to help the reader probe further.

Dom Henry is a gracious man and a deep thinker. He wisely observed that a Hebrew plural form might sometimes operate as an abstract or a collective noun – like Royalty, Divinity, Nobility, or Management. That might go some way to explaining why both singular and plural verb forms and attributives accompany *elohim* in different Biblical texts.

He pointed out that no one is ever really sure of the correct syntax with a collective noun. *"Royalty is always getting itself into trouble,"* may be correct but it sounds off. *"Royalty are always getting themselves into trouble,"* is technically incorrect but might work better in terms of understanding the sentence.

This might be enough to forgive the grammatical glitches around the word. However, this explanation only serves to re-emphasize that in *elohim* we are not looking at a simple singular. A plurality is still implied. So I am going to stick my neck out and say that the presence of the plural form *elohim* clues us that the sources of Genesis emanate from a worldview that believed in plural *elohim*.

So how should we translate it? Turning to etymology, the root meaning of *elohim* is either *"powers"* or *"powerful ones."*

Some commentators argue that *"powers"* must refer to the superlative attributes of the Almighty. But just as a phrase like *"the powers that be"* evokes in our minds a plurality of people who hold power so we can read *elohim* as indicating plural entities – *"powerful ones."* This reading of *elohim* as *"Powerful Ones"* is also more consistent with the way in which – *im* pluralizations work in any other context. For instance, in Hebrew a *kruv* is a cherub. *Kruvim* are many cherubs, not the various superlative qualities of cherub-kind.

Reading plural *elohim* as plural entities also makes much better sense of phrases indicating plural description and plural behavior. I mean phrases such as *"let us make..." "in the image and likeness of ourselves..." "humans have become like one of us..." "let us go down and confuse their language..."*

In such moments we are genuinely eavesdropping on conversations among the Powerful Ones.

I would argue that an earlier mythology which spoke of a heavenly council – a group of *elohim* – has been redacted by J (and colleagues) to harmonize the *elohim* stories with the stories of Yahweh. However, rather than create a harmony, the redactors have sown dissonance and moral confusion into our reading of the Bible ever since.

Now I think it's fair to say, devout Jewish and Christian believers – myself included – have often struggled with the idea that the book of Genesis might be an altered version of someone else's book. Theologians have often contorted themselves to avoid the obvious questions thrown up by these plural forms. Certainly, as I made my way through the commentaries and books of theology adorning my bookshelves, I found that my questions were met with no more than a meagre sentence or two before the authors would anxiously hurry away onto more confident ground.

One day I turned to a voluminous work with the modest title,

The Theology of the Old Testament, published by T & T Clark, edited by S.D.F.Salmond. In it the theologian A.B.Davidson writes this:

"In contrast to man, angels belong to the class of elohim…It might be interesting to contemplate the question of how the same name 'elohim' came to be applied to God and this class of beings."

Agreed! Good question! Then he goes on.

"Perhaps we should be satisfied with the general explanation, that the name, meaning 'powers' is applied from the standpoint of men to all that is above man, to the region lying above him. Though the same name is given [i.e. to angels and God] the two are never confounded in Scripture."

Perhaps we should be satisfied with the general explanation? I wasn't satisfied. The circle had not been squared. It fudged the question of how a unique, peer-less, almighty God could in any way be regarded and named as one of a group. The more I was served up that kind of non-explanation, the more I felt that I needed to keep asking.

What would happen if we were finally to concede that *elohim* implies plural beings? And what happens when we translate the Genesis accounts that way? Of course, the story changes. But what made my mouth fall open as I went about the exercise is that the change that results is far from random. I found that it is like brushing revealer over invisible ink, because what surfaces, previously hidden in plain sight, within the familiar verses of the Bible, is the unmistakable thread of an even older narrative. It's a narrative that changes our understanding of what the Bible is and who God is. More dramatically still it totally rewrites our understanding of who and what human beings are and where we have come from. Opening my eyes to a *"plural Genesis"* was like releasing the brake on a rollercoaster and I had no way of knowing where it would take me.

CHAPTER TWO

THE HUMANS AND THE OTHERS

When Neo takes the red pill in the movie *The Matrix* he wakes up to a new world. Everything he thought he knew turns out to have been an illusion and Neo must now find his feet in a new reality and a whole new world of possibilities. Going through the portal of the *elohim* word into the primal sources of Genesis was just like that. As I began to explore this new territory my understanding of who God is, what his role in the world is and where human beings came from all began to shift and take a new shape.

The advantage of being in traction meant that it would be some time before I was back in the pulpit. In a way I was grateful for that as it gave me the time to process my fundamental questions.

I took some time to take a closer look at all the names in the stories of beginnings. Before Adam was a name, it, too, was a word. It means *"of the earth"* because he was formed from the earth. We are so familiar with Adam as a name that it is hard for us to conceive of *adam* as anything other than a name. But it is a word. And in English we have two very close equivalents for it.

Human = the *humus-thing* (*humus* means *earth* or *soil.*)

Earthling = the-earth-thing (*earth* means *soil* or, in today's thought, *the planet.*)

That makes the stories of *Adam* stories about *the earthling* or *earthlings*.

Similarly, Eve carried a meaning before it became familiar as a name. It means *"The Living"*.

The stories of Eve are therefore the stories of the living.

Touring the pages of Genesis, I could see the sense of that reading. There's a clarity and depth those meanings bring to the theologies of Judaism and orthodox Christianity that the current version of Genesis has been crafted to teach.

To give *elohim* a similar dust-off, I needed to go all the way back to the beginning and start all over again! Locked down in my shipping crate seclusion, in the leafy corner at the end of our drive, that's exactly what I did. As I read, an unfamiliar story began to surface. Though new to me, it was in reality an old, old story, from a time long ago and a land far away. It goes like this:

In the beginning the earth is shrouded in darkness, empty and barren and covered in water. Now we see spirit-beings, the Powerful Ones hovering above the earth, orbiting the watery surface. Now we see light, manifesting sun, moon, and stars.

When we read this, we think of the universe as we know it and can hardly imagine what kind of powers or technologies or trans-dimensional manipulations would be needed for these Powerful Ones to be in the business of star-formation or planet-building. However, we probably need to note that the original tellers of the tale would probably have pictured a more-or-less flat disc of terrain, covered by a huge dome or vault, like a child's snow-globe, or Terry Pratchett's *Discworld*, or the town of Chester's Mill in *Under the Dome* or the giant studio in *The Truman Show*. Perhaps we are observing something more fathomable than the manufacture of stars and planets. Could it be that they are merely terraforming a primordial soup of land and sea?

If they are planet-formers then the Powerful Ones are beings who exist in the heavenly realm, yet who can project themselves and function with incredible power in the physical realm of time and space. If we think of ourselves as beings moving in the four dimensions of space and time, then perhaps we might describe the Powerful Ones who arrive from beyond what looks like the beginning of time as multi-dimensional or trans-dimensional entities. If, on the other hand, they are merely terra-formers, even that is no small assignment. They would still be entities with powers on a superhuman scale, to say the least.

In whichever we may conceive of them, the story invites us to recognize a power well beyond anything we are familiar with. To understate it, they are clearly different to us! For simplicity, I will stick to the root meaning of the word *"elohim"* and refer to these entities as the *"Powerful Ones."*

Now the Powerful Ones divide the waters to create saltwater seas, freshwater rivers and habitable land. Vegetation, fish, birds and animals now fill the air, land and seas. Within a plain called Eden the Powerful Ones create an enclosed zone and fill it with animal and botanical life. In the soil of the plain lie precious mineral deposits, including high grade gold. This valuable land lies close to four major rivers including the Tigris and the Euphrates.

Using the elements of the earth to make a clay, the Powerful Ones now fashion human beings to look like their makers. The bodies lie silent and motionless until the Powerful Ones breathe spirit into them.

Once animated the new human males are put to work in the enclosed zone. They eat a vegan diet and live a subsistence life in harmony with the animals.

Gradually the Powerful Ones notice that the human males are depressed. All the other animal species are male and female. The humans need such companionship too. So the Powerful Ones now generate a female of the species from the body of the male. As male and female, human society is now poised to begin its journey.

Then one day the humans find themselves in conversation with an entity known as The Snake.

This is intriguing. Thousands of years later, realizing that this is a plot-twist that needs some serious explaining, the author of the New Testament book of Revelation correlates The Snake with some other mythological figures: *"The great dragon [is] the serpent from the beginning times, called 'devil' or 'The Accuser'."* At the other end of the Biblical timeline, one of the oldest portions of the canon, *Job 1.6,* identifies the accuser as a spirit-being who is among the *benei-elohim*. The Snake is *elohim-kind* – one of a number who confer and operate together as members of the heavenly council. He is one of the Powerful Ones!

The Snake shows the humans how they can achieve a higher level of consciousness. The change will raise their understanding and self-awareness and improve the quality of their life. Not surprisingly, the humans accept his offer. It is a collaboration in which the female of the species takes the lead. This upgrade brings with it moral conscience and sexuality. Now self-aware, the humans begin to wear clothes. Gender roles begin to emerge, along with the first childbirths. It is the beginning of human society. But trouble lies ahead for the humans as a deep and long-lasting conflict begins to foment among their makers.

Diverse Biblical texts speak darkly of an all-out war among the Powerful Ones. This reveals that the Heavenly Council is not quite as heavenly as we might assume. Evidently it is not merely an assembly of avatars or agents for the will of the Almighty. The fact that the accusing spirit of Job 1 and the deceiving spirit of I Kings 22 figure on the council's line-up shows us that the Powerful Ones each have their own wills and their own agendas. This heavenly council is far from an angelic choir. Because of this I would suggest that *"heavenly council"* is a misleading description of this intriguing body. To my mind, *"heavenly"* implies divine, lovely, beautiful, and tranquil. So I will refer to it simply as *"The Council."*

If *elohim* refers to the diverse and sometimes fractious members of the Council then translating it as a single entity called *"God"* can only produce an incoherent picture. Sure enough, reading elohim in the singular as *"God"* does indeed throw up some wobbly questions. How could an all-wise *"God"* fail to anticipate the human males' need for female companionship? How could a morally good *"God"* not desire human beings to be morally aware? If the humans were not morally aware how could *"God"* hold them culpable for making a wrong decision? Why would their gaining moral awareness need to be punished by *"God"* with lifelong hard labor, painful childbirth and finally death? How could *"God"* not anticipate that The Snake would create a problem? Why would *"God"* even think to create such an entity? How could *"God"* fail to anticipate obvious eventualities to the extent that he would come to regret having made humans in the first place? The singular translation presents us with a *"God"* who appears wrong-footed, unpredictable and cruel. This is exactly the kind of moral distortion I was talking about before.

With a plural *elohim* the stop and start, push and pull of the *elohim*'s actions turns out to be not the vacillations of a double-minded deity. Rather we are seeing conflicting agendas among the Powerful Ones. It's about a falling out among the leadership of the Council.

Throughout the Bible, various texts refer to a full-blown war dividing The Council over struggles for pre-eminence and the progress of the human beings. Those of The Council who side with The Snake are outnumbered two to one and are exiled on Earth. If these conflicts are all of a piece, then this warfare and its outcome are neatly symbolized within the Genesis narrative itself:

As punishment for breaking rank and effecting this unauthorized upgrade to the humans, The Snake is exiled on Earth to eat the other Powerful Ones' dust!

The male and female humans now face the consequences of their new condition. Until now they have shared the enclosed area with the Powerful Ones, communicating face to face, with all their needs provided for. But now the enclosed zone is to be purged, with the humans locked out to fend for themselves in the wild, untamed country of Eden. There the humans begin to produce children. The land of Eden has not been prepared or cultivated and the men now have to work hard just to provide

for their families.

Now that they are barred from the enclosed area, the humans no longer enjoy access to the healing botanicals which before had cured every injury and ailment. Denied such cures the humans begin to die.

The first death, however, is a violent one – an assault by a farmer called Cain. When the Powerful Ones learn of it, the killer is banished from the region of Eden where the Powerful Ones reside. Contemplating his fate, Cain faces his fear of the people who live in the region of Nod where he settles.

As the human population continues to expand, the people encounter other Powerful Ones *(benei-elohim / elohim-kind)* **who begin abducting human females and having children with them. Their progeny are the giants of legend.**

This strange scenario of god-like beings taking human wives would seem a bizarre plot twist even for a twenty-first century movie – let alone an account purporting to explain how we got here. Yet it is a story retold in the mythologies of countless cultures around the world. Each has found a mode of preserving the same narrative from prehistory until today. It repeats in Hindu mythology and in Graeco-Roman mythology. It's part of the sagas of Norse and Celtic lore. The ancient mythologies of Mesopotamia recount it in great detail. And in all the mythologies the Powerful Ones' intercourse with humans produces demi-gods and titans – "men and women of legend."

Some of these hybrid people have become familiar to us through the literature of Greek mythology. The Greek "men of legend" include Achilles, Aeneas, Heracles, Perseus, Theseus, Orpheus and Helen of Troy – to name just a few. In all there are twenty-eight hybrid people named in the Greek panoply. Hindu lore names seventeen. There are a couple in Norse mythology and one in Celtic mythology.

As human society grows and becomes harder to manage, the Council decides that the human beings are living too long and

takes a decision to limit the humans' lifespans. Troubling news of abductions of human females and resultant hybridization also reaches the ears of the Council. This is not what they had intended. After a period of intense debate, a final solution is agreed. Earth will be wiped clean of the human menace by means of a massive flood.

But not all the Powerful Ones are agreed. So it is that a message is brought to a man called Noah. Along with a warning about the impending flood, Noah is given instructions to construct a rescue vessel and seal his family inside, together with a stock of plants and animals to reseed flora and fauna and re-boot the human population with the DNA of his extended family. Then comes the deluge.

Indigenous mythologies from all around the world tell the tale of a great flood and the re-population of Earth. It's in the folklore of China, Peru, Korea, Malaysia, Ireland, Southern Iraq, Northern Iraq/Southeastern Turkey, Hawaii, Finland, Polynesia and the Philippines. It's in the oral tradition of First Nation Americans and in the Australian Aboriginal creation story.

Three times Noah releases a bird to look for land. When the third does not return he and his family know that land has resurfaced. After the flood abates the vessel runs aground on a mountain. Now the animals re-emerge, along with Noah and his family who make animal sacrifices to thank the Powerful Ones for warning and saving them.

In the aftermath of the flood the Powerful Ones speak of the regrettable violence of the deluge and vow never to do such a thing again. They affirm a standard of accountability for murder and genocide. Even the animals must now account for any human lives they take. On instruction from the Powerful Ones, the humans now add fish and meat to their diet. Humanity will now rebuild itself with the gene-pool of five families – the extended family of Noah.

As human society re-expands the people migrate east of

Eden. When the people reach the Shinar Plain, Mesopotamia – the home of Sumeria and Babylonia – they settle and build something incredible.

This geographical detail is noteworthy because Mesopotamia is generally acknowledged as the cradle of modern civilization – equipped with the whole package of agricultural systems, cityscapes with streets, plumbing, and the beginnings of metal technology. Here hours and minutes were invented. Mathematics was pioneered here. So was money and banking. It also was home to the world's first writing. With writing came law-making, record-keeping, money, banking and literature. All these things started in Babylonia – all of a sudden. Then rapidly this package began springing up all round the planet, as if the concentration of expertise in Mesopotamia had been blown by the four winds to self-replicate all around the world.

What the people now construct in Babylon is a tower – a gateway between the people and the Powerful Ones, a means of reaching the heavens from Earth.

If El is short for elohim then Babel means a *Gateway to* or *for The Powerful Ones.*

However, when members of The Council make a visit to observe the project they are deeply disturbed, fearing that the humans have become too capable. Another brutal response is called for. The Powerful Ones use their great powers to destroy the city state and arrest the development of human civilization. They do this by taking from the humans the language which has united them, confusing their ability to speak so that the humans are no longer able to communicate with one another and operate as a single society.

Nothing happens after that until in **Genesis 12:1** we encounter the great patriarch, Abraham in a world that looks pretty much like ours. He has to negotiate a world of nomadic tribes, city states, and the beginnings of empires. It is a world of trade, wars and peace treaties. The religious and cultural environment is one of diversity and negotiation. It is, more-or-less, the world as we know it. But the journey by which we have arrived has been a bizarre one.

My discoveries were challenging at many levels and, as you'll appreciate, they did not present likely material for preaching any time soon! The paradigm shift was challenging enough for me as I pored over my interlinears, lexicons and commentaries for weeks on end. I could hardly expect my hearers to make the imaginative leap I had in the space of a couple of twenty-minute sermons.

However, although the story in bold may sound bizarre on a first hearing, for some – and maybe for you – it will ring more familiar notes. If you're at all acquainted with the mythologies that flowed from ancient Mesopotamia, you will quickly recognize that my retelling of Genesis is far from a piece of random fiction. It is a story that already exists. It resides in the oldest depository of writing known to humanity, etched indelibly in cuneiform script on the dark-stone tablets of the ancient Sumerians, Babylonians, Akkadians and Assyrians.

At one level this literary overlap shouldn't really come as any great surprise. Abraham and Sarah – the progenitors of the Judaeo-Christian tradition – grew up in Sumeria. They left as adults from Ur of the Chaldees. Abraham and Sarah's whole education was a Sumerian one. So within the Bible that was ultimately to issue from their family's faith we should fully expect to find the remnants of their culture of origin. Yet, once highlighted, these literary overlaps do far more than simply add a gloss or two to our familiar story of beginnings. They change the whole picture.

That's because in its current form, the creation account of Genesis stands as a kind of antithesis to the Sumerian narratives and their Babylonian, Akkadian and Assyrian spin-offs. The Book of Genesis seems to say, *"Never mind those old stories here's what really happened!"* But the moment we re-translate *elohim* as a plural, or as a collective noun, Genesis does an incredible *volte-face.* Instead of critiquing the Sumerian account, Plural Genesis actually confirms it, story after story.

This was quite a shift for me. Somehow, I had always pictured the Bible as if it were the truth against the world. It turns out the truth is even stranger.

Incredibly I would have known none of this had it not been for an 11-year-old boy from Western Iraq – a boy by the name of Eylo, who someone had persuaded to dangle by a rope over a 100 meter cliff.

CHAPTER THREE

STRANGE! I'VE SEEN THAT FACE BEFORE

This was a difficult rock wall to climb. As a fit 11-year-old Eylo had the agility, with hands and feet small and strong enough to get a grip on the meagre hand and footholds of the cliff face. A hundred meters below, the rocks of the desert floor gave him reason not to slip. Above him soldiers ballasted the rope strapped carefully around Eylo's waist. The soldiers were in turn supervised by a bald-headed gentleman sporting a fine military moustache – a fashion-must in the armies of Britain and the East India Tea Company. It was 1835 and the mustached man was an Englishman by the name of Henry Rawlinson.

Rawlinson's official reason for being in this particular part of Western Iraq was to help the Shah of Iran train his troops. But what he was doing on this day in Behistun, dangling Eylo over a cliff, was an attempt to reach an inscription carved into the stone of the cliff face. There was a reason the inscription was worth the effort.

The Behistun inscription was a royal proclamation written in cuneiform script. Cuneiform is the world's most ancient form of writing, dating back five and a half millennia. It was used until the first century AD when it fell into obscurity. In the 1500s it was re-discovered, found inscribed upon tens of thousands of clay tablets buried in various archaeological sites around the territories of ancient Mesopotamia.

However, the secrets of the re-discovered cuneiform tablets were to remain locked up for the best part of four centuries. Kept under lock and key in various imperial museums, the tablets held on to their mysteries. Western scholars who analyzed them somehow failed to recognize the wedge-shaped markings of cuneiform, etched by blunt reeds into the clay tablets, as being any form of script. The scholars took the markings as purely decorative. So it was that until the 1800s the cuneiform tablets remained mute, a bank of hidden treasure waiting for someone to find the key.

The Behistun Inscription was that key.

What Eylo carefully copied for Henry Rawlinson as he dangled on that rope in 1835 was a royal proclamation in three known languages, Persian, Elamite and Akkadian, all expressed in cuneiform script. It was the three known languages that provided translators with the key to unlocking cuneiform for a new world, hungry for its secrets. Just like the Rosetta Stone was for Egyptian hieroglyphics, so the Behistun Inscription was the codebreaker that gave a new era access to the heritage of a long-lost civilization. Suddenly we had a window onto a forgotten world.

Now at this point I should come clean and tell you that I made Eylo's name up. That's because unfortunately, in all the excitement of the day, Henry Rawlinson neglected to note down the name of the boy who had made such an effort for world history. Eylo is a Kurdish name meaning Brave Eagle. So I thought I would honor the brave young abseiler with that name.

As the more than 200,000 tablets that had been unearthed began to be translated, we found we were gazing through portals into all kinds of aspects of Sumerian, Akkadian, Assyrian and Babylonian life. The tablets recorded legal notes, business agreements, banking arrangements, inventories, and shopping lists. But what really caught people's attention were the stories of beginnings. On the one hand the explanations of human beginnings etched in cuneiform appeared quite foreign and bizarre – clearly the product of an unfamiliar thought-world. On the other hand, some intriguingly familiar motifs began to reveal themselves.

It is an uncanny exercise, even today, to read the transcripts and translations of these ancient accounts. It is like stumbling across a forgotten photo album of some distant relatives you hardly knew about. At first you feel like you're spying on something faintly comical. You laugh at the out-of-date fashions and the bad hair. You are spying on someone else's life, viewing snapshots of places and events that must have been meaningful to the original viewers, but which are coldly irrelevant to you. Then, from out of the blue, a face appears that you know. And then another. And another. Places and events that are part of your own family's story. What are the members of your family doing in this family's pictures? What's the connection? This is when you realize there must be a whole lot of stories about your own family that you don't know. It gives you a funny feeling and leaves you wondering how much you really know about your own family.

As I sat immersed in papers and reference books, translations and lexicons, the stone snapshots of the Mesopotamian tablets began to give me that same funny feeling. I began to wonder what I might not know about my own religious heritage.

It was the same funny feeling that disturbed many good religious people back in the nineteenth century as these forgotten stories re-surfaced, all predating the familiar canon of Biblical accounts. In the cuneiform glyphs transliterating Sumerian, Babylonian and Assyrian accounts some familiar motifs repeat. The stories of beginnings include the Sumerian *Epic of Gilgamesh, The Eridu Genesis, The Creation of Humankind, The debate between Grain and Sheep;* the *Babylonian and Assyrian, The Seven Tablets of Creation and Enuma Elish (tr When in the heights.)*

All are found on tablets dating from around 3000BC to 1100BC. But their colophons (the ancient authors' blurbs) tell us that even these ancient accounts are just the re-telling of much older accounts. At first they make for quite dense reading. I wasn't reading modern prose. I felt that I was trespassing in the abandoned halls of a deeply foreign culture. But as I continued to read, I found myself gravitating to the gallery of familiar faces, places and events. I wanted to know what the tablets had to say about them. This is what I found…

(In this summary account I have favored the names used in the Sumerian versions of the stories. This is because Babylonian and Assyrian societies came later and are daughter cultures of the Sumerian culture. So although the oldest surviving copies of some of the stories that follow are found in Babylonian and Assyrian cuneiforms, their versions were developed from the older literature of their parent culture – that of the Sumerians. Hence the Sumerian names are the older, original names for these characters.)

Before the creation of anything is Anu the source of all things. Earth lies in darkness. The salt waters and fresh waters, both birthed by Anu, must now be separated. These two bodies of water correspond to two powerful entities Abzu (male) **and Tiamat** (female.) **Abzu and Tiamat bring forth other powerful beings who will collaborate in creating the world, beginning with light.**

There comes a time when the powerful beings fight among themselves for power and pre-eminence. The order of the Council must be established, and a CEO recognized. Enlil is installed in power. (Enlil means *"Air-Lord")*

The Sumerian word for *"god"* or *"gods"* is a glyph that indicates the sky. To get as near to the original associations as we can I will be referencing them as Sky People.

In the beginning Earth is empty and barren but the Sky People create an enclosed zone within a region called Eden and fill it with animal and botanical life. The region the Sky People have chosen is close to fresh water, including the rivers Tigris and Euphrates.

Enlil now confers with his brother and First Officer, Enki. (Enki means *"Earth-Lord.")* **Together they decide to create human beings. Enki will now oversee the human experiment.**

As he plans, Enki collaborates with Namma – the original *"mama"*.

One of the Sky People defeated in the wars by Enki is Qingu. His DNA, carried in the medium of his blood, is now put into clay to generate the human beings. In the nursery Enki partners with Ninhursag – the original *"nurse"* who nurtures and nourishes the new humans into being.

The very first human males live in the wild. They eat a vegan diet and live a subsistence life in harmony with the animals.

The Sky People now decide to modify the humans and take humanity to a new level. A female from among the Sky People introduces the wild man to bread and beer and brings the humans into the cultivated zone where they are taught how to live *"civilized"* and how to cultivate crops.

The Sky People then upgrade the humans by creating the female of the species. This upgrade to human society brings with it a new self-awareness and sexuality and a higher level of consciousness. The humans now begin to wear clothes.

These details, drawn from different stories among the cuneiform tablets really caught my eye. They rang some bells. For instance, I had always noted that the writer of Genesis expects us to feel a delight and empathy when we read of Adam's original state of innocence, naked and living among nature, in harmony with the animals. That empathy reflects in so many of the world's fables of human beings living in mythological times as one with nature and in a world where we and the animals could talk. I see this same deep feeling expressed by our children's delight in sharing their lives with toy animals and with pets. Might the deep feeling really be the vestige of a deep memory?

It is a special part of Australian Aboriginal culture for males to go walkabout. It is part of their initiation into adulthood and becomes a vital tool for balancing and re-centering their lives. It is a ritual in which the young men respond to the urge to go out into the bush, to live alone in harmony with nature. Might this cultural practice, resonant with feeling, also be the expression of a deep sense of memory?

Then there is the resonant Western story, told by Edgar Rice-Burroughs, in which civilized Jane finds an innocent Tarzan living wild and naked in the African jungle, in perfect harmony with the animals. She awakens his sexuality, teaches Tarzan to speak, civilizes the wild man and brings him to the city. It's the same story.

Rudyard Kipling tells the story of Mowgli, living innocent and naked in harmony with the animals in the jungles of India. He too is awakened and civilized by a beautiful girl who takes him to the village where he can learn to be human.

Is the repetition and resonance of these stories no more than a coincidence? They repeat the same themes of awakening, educating, and civilizing the wild man and bringing him into the village or city.

As I read on, I was to understand that in the cuneiform's telling there was a less altruistic motive behind upgrading the humans and bringing them into the enclosed zone.

Once in the city the humans learn they have been created to relieve the Sky People of menial work, tending the crops and keeping the Sky People supplied with food. The humans are taught to harvest the crops and bring them to the Sky People as sacrifices. For a while this pattern works well, freeing the Sky People to attend to other more specialized things.

Enlil now divides the Sky People into two groups. Three hundred are assigned to look after the Earthly zone known as the Abzu. Another three hundred will become Observers who will watch and protect the Earth from the sky where Enlil has created stations for them among the stars. The main station among the stars is called Nbiru.

Now male and female, the human beings begin to procreate. As they multiply and produce families the human population becomes noisier and more difficult to manage within the city.

The Council agrees that the humans are living too long and decide to limit the humans' lifespan.

Troubling reports have reached the ears of the Council. Human females have been abducted by people who are part-human and part-Sky People. These new levels of hybridization create a sharp disagreement among the Council. Enlil insists on a final solution whereby the Earth will be purged of humanity by means of a massive flood.

But not all the Sky People are agreed. Enlil's First Officer, **Enki, has overseen the development of the humans and refuses to give up on the experiment. Enlil now goes in secret to warn one of his most trusted humans about the impending flood. The man's name is Ziusudra, and he is the king of Suruppak.**

Ziusudra is the man's Sumerian name. He is Utnapishtim in the Babylonian version. Atrahasis is his Akkadian name.

Ziusudra receives Enki's instruction as to how to construct a rescue vessel. He is to seal his family inside, together with a stock of plants and animals to reseed the various species and re-boot the human race with the DNA of his extended family. Then comes the deluge.

Three times Ziusudra releases a bird to look for land. When the third bird does not return Ziusudra and his family know that land has resurfaced. After the flood abates the vessel runs aground on a mountain. Once on dry land, the animals re-emerge along with Ziusudra and his family who make animal sacrifices to thank Enki for warning and saving them.

In the aftermath of the deluge the Council confers over the regrettable violence of the flood. The Council condemns indiscriminate killing and affirms personal accountability for wrongdoing. Some of the Sky People suggest that animals taking the surplus human lives would have been preferable to the horror of the deluge.

(In a much later version of the story, recorded in Greek by the Babylonian priest Berossus, a voice then instructs Ziusudra to migrate East and establish a city in Babylonia.)

In Babylon the Sky People engineer an incredible structure. A grand opening celebration is held and a Council of Fifty is set aside to govern operations. From within the new structure seven technical experts dispatch the three hundred observers to their stations in the sky.

This is a particularly intriguing detail because in the parallel Biblical account the name of the structure, *"Babel"* translates as *"gateway for the elohim" – if el is short for elohim.* Genesis 11 specifies that Babel has been constructed as a means of reaching the heavens. So here the cuneiform and Biblical accounts serve to amplify and finesse each other. Today we would call the stations in the sky *"space stations"* and the sending structure a *"stargate."*

Admittedly the touchpoints I have highlighted are only a thread within the Sumerian story. It would be fair to say that the Sky People get up to a lot of activity in the cuneiform tablets that is not detailed in Genesis. On the other hand, a great deal of the extra activity of the Sky People could be accurately summarized in terms of a number of Biblical texts.

Passages such as *Isaiah 27:1; 51:9–10, Psalms 89:9–12; 74:12–14; Job 9:13–14; 26:12–13, Ezekiel 28, Luke 10, Revelation12:7–17,* all refer to battles which pitch God against various mythological entities such as Leviathan, Rahab, Behemoth and the Dragon. Some writers have suggested that these are symbolic references to the enemies of Israel – Egypt, Babylonia, Assyria, and even the Red Sea. Or, as German theologian Hermann Gunkel suspected, they may be images and epithets of a more ancient lineage.

The New Testament book of Revelation neatly summarizes the extra cuneiform storylines in six Greek words: *"And there was war in heaven."* The war is around struggles for pre-eminence and the progress of the human race.

Could these parallels really be no more than a coincidence? It is undeniable that the strange story of the cuneiforms has a great deal in common with a Plural Genesis (i.e. Genesis translated with a plural *elohim*). What does that parallel mean? Does it mean something or am I wrong and *elohim* is really just another more general name for Yahweh?

To settle questions like these is there any kind of proof – a smoking gun to show that this is that – and that before the flood human beings really did interact with other intelligent entities – *"Sky People"* as I've called them in the tablets and *"Powerful Ones"* as I've called them in Genesis?

Once you have seen the similarities it is impossible to just walk away from them. Either Genesis has retold the cuneiform stories or Genesis and the cuneiform tablets are, together, recalling ancient narratives known to both cultures. Given that Abraham and Sarah were already mature people when they left Sumeria it would seem most logical that it is the narrative of Genesis that has recast the Sumerian account, which had also seeded the Assyrian and Babylonian versions.

To some this would seem a quite uncontroversial claim. To others, though, it is taboo because it appears to undermine the sense of divine authorship for the Bible. Even to shine a light on the parallels offends some devout believers.

Nathaniel Schmidt found this to be the case back in the day when he was Professor of Semitic Languages at Colgate University in the United States. Schmidt was a Baptist Minister who served at the university for eleven years, teaching courses in Hebrew, Arabic, Aramaic, Coptic, Syriac, and many other ancient languages. It really was a scoop for Colgate University to have such an erudite scholar on its faculty. That is until sometime in the 1890s when Schmidt began publishing papers that highlighted some of the parallels between the Biblical and Mesopotamian stories of beginnings.

By way of thanks for those particular papers Schmidt found himself tried for heresy in 1895 and fired from his position in 1896. Nathaniel Schmidt's next publication was, aptly, a study titled, *"Biblical Criticism and Theological Belief."*

Fortunately, the next turn in the road for Schmidt was a positive one as Cornell University jumped at the opportunity to gain a scholar of Schmidt's caliber. In fact, Cornell happily retained Schmidt as Professor of Semitic languages for a full 36 years. Nevertheless, Professor Schmidt's experience stands as a sobering reminder that the development of ideas is not always a smooth or easy path – especially in the world of faith. Think Martin Luther. Think Galileo!

Speaking for myself I enjoy my work as a pastor, teaching and preaching from the Bible for faith groups around the world. I love teaching pastors and theological students the time-tested principles of hermeneutics – the tools by which we interpret texts – and applying them to the Bible. How could I not enjoy the kudos of being regarded as both orthodox and thoughtful? And so as the parallels impressed themselves upon me, I was not entirely sure how ready I really felt to follow in the footsteps of forbears like Nathaniel Schmidt or my friend Vince – the guy I mentioned from that heterodox sect. For that reason I was eager to lean in to the expertise of thinkers cleverer than myself to get my bearings.

However, as I began to sound out academic friends as to their awareness of these parallels, and what they made of them, a surprising logic began to repeat among friends who were mainstream Christians. No, my academic friends did not find the parallels disturbing! In fact, some were willing to state that it was blindingly obvious that the Biblical stories of beginnings were based on the Mesopotamian stories. But if I asked the obvious next question, their faces would always fall. If I asked, *"So do you think the Mesopotamian accounts of Sky People, and how they engineered humanity, are true?"* the answer was always along the same lines. *"How can they be? The Bible's versions of the stories are true. It's what the Biblical versions are based on that's false!"*

That reasoning baffled me. The idea of a true story being developed from a false story isn't impossible but something odd happens when faith is added into the mix. Interpretation aside, the orthodox Christian believer accepts the authority of the Bible as a fundamental. Most Christian doctrinal bases denote the Bible as something like the *"supreme authority in all matters of faith and conduct."* The Church of England's articles of religion declare that the Bible comprises those writings *"of whose authority was never any doubt in the Church."*

Once the parallels between the biblical and Mesopotamian mythologies are acknowledged, the position of faith then says that in order to view the Bible's accounts as true or *"authoritative"* one must view the texts on which they were based to be false.

As reassuring as it may be to go with the flow of orthodox belief, I just couldn't get my head into that kind of logic. Surely that's like a historical figure being the descendant of a fictitious character? As to the idea that, thousands of years after their creation, the Mesopotamian account of origins could be *"corrected"* by a someone who somehow knew better; it seemed to me this was an idea without roots in anything other than a pure assertion of faith.

If we can accept that in Genesis we have a narrative that has drawn from the earlier Sumerian source – as relayed in the Mesopotamian tablets – then at some point we have to square up to the implications of the parallels between the Sumerian Council of Sky People and the Bible's Council of Powerful Ones. Is it that the two sources draw from the same well of ancient memory? How can we be sure? Where will we find our smoking gun?

For an illustration, imagine that your friend John owns a blue Toyota Fortuner. It has white leather seats, a leather dashboard glove and a badge, saying *"John and Judy."* It's a car you're very familiar with, as you have ridden in it many times.

Imagine that one day I take you for a ride in my new car. It's a Toyota Fortuner. You notice that, just like John and Judy's, mine has white leather seats and a leather dashboard glove. Little details here and there convince you that I have stolen John's car. So you challenge me. *"Have you pinched my mate John's car?"*

"Of course not!" I say, all innocently. *"Mine may look similar. Yes it's got white leather seats and a leather dashboard glove – exactly like John's. But my Fortuner happens to be orange and the name on the badge is Paul!"*

Am I telling the truth? Are they two different cars or have I really just pinched John's car, had it resprayed and rebadged it?

To settle the matter beyond doubt you just need to locate the car's unique fingerprint, the vehicle identification number, etched into the metal. That's the smoking gun you would need to find.

If the Powerful Ones of the Bible and the Sky People of the cuneiform tablets are the same, and not similar by coincidence, then presumably Yahweh is a separate entity whose self-revelation has been interwoven with the stories of the Powerful Ones. For our smoking gun we need look no further than the book of Joshua 24:14

In this passage Joshua gives a speech to persuade the people of Israel to entrust themselves to the entity who revealed himself to Moses in the desert as Yahweh. Joshua has succeeded Moses as leader following Moses' death. He addresses the people in these words:

"Now, therefore, follow Yahweh and serve him in sincerity and truth. Put away the elohim (the Powerful Ones) whom your ancestors served on the other side of the river and in Egypt, and serve Yahweh... Today make up your minds whom you mean to serve; the Powerful Ones whom your ancestors served beyond the river, or the Powerful Ones of the Amorites in whose country you now live. But as for me and my house we will serve Yahweh."

The time in Egypt speaks of the Israelites' time enslaved within a foreign religious culture before Yahweh revealed himself to Moses. *"On the other side of the river,"* refers to the religious culture of Abraham and Sarah's roots in Sumerian

Mesopotamia. Joshua 24:2 clarifies this:

"Since a time before memory your ancestors, right up until Terah, the father of Abraham and Nahor, lived on the other side of the river and served Powerful Ones (i.e. other (plural) *elohim.)*

Joshua calls his people to reject the Egyptian and Sumerian *elohim* – *"tear them off"* is his phrase – and give all their allegiance to Yahweh.

Here is an instance when *elohim* and Yahweh clearly cannot be equated. Here *elohim* is a plurality of Powerful Ones whom we are not to worship. They are the Sky People of Abraham's Sumerian heritage, whose stories are told in the cuneiform tablets. Joshua calls the people in God's name to disregard them, reject them and cut them off!

For me this is the smoking gun. Every point of correlation between the Sky People and the Powerful Ones in their respective storylines constitutes another digit in that vehicle identification number. Joshua's speech unambiguously connects the two mythologies.

Escaping Sumeria was an escape from the Sky People. In the same way escaping Egypt was an escape from subjection to the ancient Egyptian regime – including the influence of their Powerful Ones. Now aided and led by Yahweh, the children of Israel could be free.

If the Sumerian and Egyptian cultures and the plural *elohim* narrative all recall subjection of human beings to Sky People or Powerful Ones who are not God, then we have to ask ourselves who or what the Sky People or Powerful Ones are exactly? What kinds of entities would be present and interacting in the prehistory of the human race?

One possible clue can be found among the shopping lists, legal documents and inventories of the cuneiform tablets. The King's List refers to tablets recording the reigns of Sumerian kings going back through history to beyond the time of the deluge.

These lists would make for unremarkable reading except for one detail – the dates. Among the more recent reigns recorded we read of incumbencies of anything from 6 to 36 years. But as we go further back through history something anomalous appears. Out of the blue we find a dynasty that lasted 24,510 years, 3 months, and 3½ days. That's a very precise number. It is a precision we would expect of the Sumerian culture – the culture that pioneered the mathematical model that gave us the second. However, it's the next sentence that blows the picture open. The 24,510 year, 3 month 3½ day dynasty was divided over no more than 23 kings. That's a mean tenure of more than a thousand years each! Those kings would have been older than Methuselah (the longest lifespan recorded in the Bible!) The dynasty concluded by the great flood was one of 241,200 years divided by 8 kings. That's an average reign of 30,000 years each!

Comparing the longevity of these impossibly long-lived kings with a normal human lifespan is like comparing a human lifetime with that of an ant! So what is this? It presents as nothing more than a dry, clerical catalogue of dates recorded for posterity, chiseled into stone tablets without any comment, fable, poem, hagiography or song to lend these dates any other layer of meaning, a record that just happens to include kingly reigns ranging from 6 years to 36,000.

Of course it's tempting to disregard these anomalous lifespans as if the Sumerians were either rather vague or just a bit careless in their concept of time. Except that these were not people who were vague about time! Today we divide time precisely into seconds, minutes and hours because the Sumerians invented a model for dividing time and space which was to provide us with seconds and degrees. Today we are able to use the precision of mathematics to perform our calculations because the Sumerians invented it. Perhaps we ought to give their culture a bit more credit when it comes to their understanding of time.

Similarly, it was the Sumerians who produced a script

which could transliterate countless languages and record their literature for posterity. So we should certainly give the Sumerian scribes credit for the ability to translate terms.

The Babylonians used Sumerian mathematics to measure and record in cuneiform script a timed account of a solar eclipse with such precision that today scientists have been able to use their account as a means to calculate the deceleration of the Earth's rotation in the time since. All in all we have to rule out any idea that we are reading a litany of lazy math from people to whom time meant nothing.

The Sumerians' connection with the stars may give another layer of meaning to these dates. Some researchers note correlations between the timelines of the Kings and the period of Earth's precession – the period of its wobble – resulting in a cycle of equinoxes. Other scholars note that some kings have been listed consecutively who, according to Sumerian history, ruled as contemporaries on the thrones of different capital cities. So there may be more going on with these dates than immediately meets the eye.

Another detail worth noting is that the Kings List asserts that it has carefully catalogued reigns *"from the time the Kingship descended from the heavens,"* to the time of the flood, and then *"once more the kingship descended from the heavens."*

That concept by which monarchy begins with *"gods"* who later bequeath the regency over to human kings and queens is not the monopoly of Sumerian culture. The Egyptians had it too.

Within this enigmatic succession, transitioning from Sky People to humans, there is a famous crossover king. King Gilgamesh – a hybrid of human and Sky People. He is the hero of the oldest written story of the world, *"The Epic of Gilgamesh."*

So when in April 2003 a German-led research team announced it had found King Gilgamesh's tomb it was an incredible moment in the history of religion and archaeology. Not much more than a month after the first US invasion of Iraq under George H.W.

Bush a research team led by Jorg Fassbinder, of the Bavarian department of Historical Monuments in Munich, announced the incredible discovery they had made. With scarcely contained excitement Fassbinder spoke to the BBC and said:

"The most surprising thing was that we found structures already described by [the Epic of] Gilgamesh…We covered more than 100 hectares. We have found garden structures and field structures as described in the epic, and we found Babylonian houses…Very clearly, we can see in the canals some structures showing that flooding destroyed some houses, which means it was a highly developed system…[It was] like Venice in the desert."

"By differences in magnetization in the soil, you can look into the ground…The difference between mudbricks and sediments in the Euphrates river gives a very detailed structure…I don't want to say definitively that it was the grave of King Gilgamesh but it looks very similar to that described in the epic. We found just outside the city… in the middle of the former Euphrates River the remains of such a building as could be interpreted as a burial place."

Potentially it was one of the most significant archaeological finds…ever. Once the find had been located and cordoned off, the first Iraq war ended within a matter of days.

What a find! A truly unique opportunity to study the intersection of mythology and history by opening up the burial place of the world's oldest mythological hero. How incredible it would be to examine King Gilgamesh, revered as a demi-god. It would provide an unimaginable insight into the world of the Sumerian Kings and a key to interpreting the Kings List.

Sadly, it appears that the public investigation of the site was quickly interrupted. In 2005 Fassbinder wrote rather mournfully in the scientific journal *Dossiers – "Archeologie et Sciences des Origines."* To translate it from the French text, Fassbinder had this to say about the find:

"Contrary to what some journalists have claimed it isn't at all proven that our find corresponds with Gilgamesh's under-river tomb.

We are sorry not to be able to give a more precise idea of the results of our magnetic investigation. But since 2003 all the archaeological sites of Iraq have been under a serious and growing threat. The lack of security in the country...[and] the trafficking of...art and artefacts are causing the total and irreversible destruction of archaeological sites by looters. All...archaeological structures will be better preserved if we leave them under the ground, untouched and buried."

Buried? What a shame! I suppose if Jorg had written this update after the second Iraq invasion he could have added ISIS as a reason that the site either couldn't be investigated or had been already destroyed. How unbelievably disappointing that even in the fifteen years since its discovery, for a find so uniquely important, we haven't been able to guard it or go back for a second look.

So for the time being then, we will have to consider the Sumerian Kings List as holding on to its mystery for a little longer. As to its longer-lived kings, we can say that the list portrays beings who, though long-lived, were not immortal as we would conceive gods to be. Who were they, then?

Let's review some of the details related by the cuneiform tablets. By any reckoning, the Sky People are powerful beings. Think about the space stations, the star gate, the Sky People who come down from the heavens, terraform barren land, manipulate weather systems and genetically engineer human beings. They are long-lived but mortal. They use technology, get into politics, break rank, battle for power, argue, experiment and invent, observing, sometimes helping, at times abducting and exploiting human beings. I would suggest the tablets paint a pretty clear picture. The Sky People are what today we would call *Extra Terrestrials.*

By the time my thinking reached this point I needed to get out of my shipping crate and go and find some older, wiser people who could orient me in the world that suddenly comes into focus once you've taken the red pill of history – meaning the cuneiform tablets.

I needed to hear from some experts from outside the realm of literature, out in the tangible flesh-and-blood world of history; people who deal with empirical things – archaeologists and paleontologists – and perhaps an anthropologist or two. Perhaps I could find a proper scientist to reassure me that humans are just humans, living comfortably alone in the universe – or to brace me for the alternative. This journey was about to take me out of my seclusion, beyond the state of Victoria and out of Australia; indeed out of all my comfort zones and all around the world on a paradigm-shifting voyage of discovery. But first I needed a beer!

CHAPTER FOUR

A LOT OF FORGETTING

"Paul, do you realize that what you're saying sounds absolutely crazy!"

Brad had what some people like to call a *"gift of encouragement"* and over this particular beer he was giving it out in spades. We were enjoying a glass at my favorite watering hole at Oldstream Pass.

"A beer will be good for your leg," he said. *"Good muscle relaxant!"*

So while my glass of Heferweiser went to work on my leg I put to Brad some of my speculations.

"If you re-pluralize that word it portals you back to the earlier version of the story. In that version, in the time leading up to the flood in Genesis 6, there are at least four kinds of entity in the text: two extra-terrestrial species – elohim and ben elohim – plus humans – and a hybrid species called nephilim. And they're all plurals!"

"Same deal in the cuneiforms," I continued, *"At least three kinds of entity: one ET species – the Sky-People – then humans and some random hybrids. Same scenario!"*

Brad looked nonplussed.

"Paul, you're pinning a lot on one word. What if, back in the day, the editors who changed elohim into God somehow knew better than you and actually got it right? What if elohim just happens to look like a plural?"

"But Brad it doesn't just look like a plural. It behaves like one. Plural form, plural verbs, plural behaviors, plural agendas – plural moralities even. As soon as you re-pluralize elohim the stories begin to make sense and the anomalies stop being anomalies."

But my friend Brad remained impassive.

"Paul, you've always prided yourself in being orthodox to the bones. What's the point of that if you're going to publish this? People are going to buy this just to laugh at it! I mean, just think about it

for a minute. If what you're saying is true – that human beings were engineered by ETs – don't you think we'd remember? That's a huge bit of history for us to forget. That's a lot of forgetting! It's ridiculous. You couldn't forget something that big."

Fair point. How could an intelligent species just up and forget where it came from? Brad is a sci-fi junky – which is why I had thought he might be intrigued by what I was writing about. So I reached for a sci-fi metaphor...

"Brad, the truth is out there! Maybe sometimes the truth is so out there that we can only speak it as fiction. Think about the Tim Burton version of 'Planet of the Apes' – the one with Mark Wahlberg." This was a movie we had both seen.

"In that story something big has been forgotten. We're on a planet where apes rule and humans are the dumb slaves. If anyone were to say, 'There was a time when apes were the slaves and humans were the masters' they would be absolutely ridiculed. They'd be shut down. They would be out of a job!

"But then it turns out there is this forbidden area where the apes aren't allowed to go. The whole area is guarded by gorillas to prevent access because it is a cultural site, a holy site. It's protected. The apes call it CA-LI-MA. It's the birthplace of their god 'Semos'.

*"However when Mark Wahlberg finds it and dusts off that old word, we discover that the strange syllables CA-LI-MA are actually the remnants of some older words: '**CA**ution **LI**ve ani**MA**ls.' The words have been obscured by the sands of time.*

"It turns out that the forbidden area conceals some ancient technology, which includes an enclosed area where the live animals in question were kept. Semos was one of those animals – not a god, but an ape with higher powers because in the time when humans were the masters they had genetically modified him!

"Only the old, old ape – Charlton Heston – knows the truth. His name is Zaius and he is the guardian of the forbidden knowledge. Hidden in his own home Zaius keeps an artefact from that long-forgotten time. It's a weapon of phenomenal power. A pistol. His son-

in-law can't believe it when Zaius repeats the insanity as if it were history, 'Once apes were the slaves and humans were the masters!'

"It's the same story! The apes get genetically modified by a higher species to become a more useful slave. That's exactly what Plural Genesis and the cuneiform tablets say about us!

"In Planet of the Apes, after the higher species disappears, the story of their ape origins gets half-forgotten, half-buried with the crucial artefacts, all hidden away to keep the story forgotten.

"How the apes do their forgetting is exactly how we do our forgetting.

"The meaning of words gets obscured by the sands of time. Anomalous artefacts get hidden away in the bowels of our museums. UNESCO world heritage and UN military keep us away from our forbidden areas. And we question the sanity of anyone who repeats the unofficial story. It's the same thing! I reckon that's what the movie is all about!

"For instance look at what happened to that guy I mentioned before, at Colgate University. He got fired just for telling the Sumerian story!

"Look at what happened to John Mack. Harvard Professor. Pulitzer Prize Winner. He was Harvard's Head of Clinical Psychology. All good until the 1990s when the U.S. Military invites John to do some psych assessments of senior military personnel who had filed reports of close encounters with ETs. When he does, he begins to notice some unusual patterns. Then he probes a bit further, widening his sample base to include civil aviation personnel and others. When he brings his report back to the military he says, 'Something is going on that cannot be explained as a psychological phenomenon. These people are experiencing something that we need to look into.'

"The moment he puts that into the public domain, authorities jump into action to destroy him. He is intimidated. His job is threatened. He's ridiculed in the press. He had to bring in top legal counsel just to keep his job. It was horrendous. And the message was clear. Let others beware!

"That's how we 'forget'! If forgetting is rewarded and remembering

is punished, it's easier to forget.

"Same deal in Christianity. The classic example is Marcion in the second century. He was a bishop with churches across Asia Minor and the Mediterranean who followed his theology. His argument was that the God and Father of Jesus comes across totally differently to the angry, genociding God of the Old Testament. 'They can't both be accurate representations of God,' he said. 'If we measure claims about God against Jesus then we really have to jettison the Old Testament, because the two visions of God are simply incompatible.' Marcion believed the elohim stories were about a completely different kind of entity.

"Today we remember Marcion as a heretic. What we forget is that those who formalized the orthodox line actually agreed with ninety percent of what Marcion was saying!

"Now Origen comes along to help the orthodox bishops answer Marcion's questions. He says, 'Don't worry everybody. You don't have to take those problem stories of the Old Testament at face value. When God does something indefensible, just don't preach the plain meaning. Read it some other way. Find the moral of the story and preach on that. Or can you read it as a prophecy about Jesus or the Church? Preach it that way and you'll find there's no problem! Jesus and the Apostles never preached the plain meaning. They always drew on the other layers in the old stories. So feel free to do the same!'

"And that, more-or-less, became the orthodox line. There was a stream that leant towards more literal interpretations – it was called the Antiochene school. But if ever an orthodox preacher needed a bit of wriggle room to avoid the real difficulties in a bit of the Bible, it was Origen's threefold reading that gave the church a framework. What Origen taught is really what most preachers have done ever since, without even thinking about it.

"But when it's all boiled down, the fact is Origen and Marcion saw the exact same problem. There is a basic incompatibility between the Old and New Testaments' vision of God. Both were actually agreed on that. It's just that Origen and Marcion each found different ways of

cordoning off the problem.

"Because Origen's line became orthodoxy and Marcion got thrown out as a heretic we tend to forget all the good that was in Marcion's contribution. He gets destroyed. Excommunicated. And there's not a trace of anything he wrote. For all that Marcion still had a massive following among the early churches, who all had to learn to be ignored or insulted.

"Listened to or not, the Marcionite churches fully believed that the elohim stories are simply not God-stories. Their interpretation of Genesis was that other entities were involved in the creation of humans and of the world we know. So it's not a new idea.

"By excommunicating Marcion and labelling him as a heretic the mainstream bishops were saying, 'Brothers and Sisters, we don't have to think any further about all that. A decision has been made. Never mind that all those in his churches in Asia Minor and the Mediterranean hold to Marcion's reading…" (a church network that survived Marcion by half a millennium.) *"…No, don't worry. Let's keep it simple. If we label Marcion a heretic we don't even have to look into that!*

"They even destroyed Marcion's work so that future generations of believers wouldn't be in danger of stumbling across this alternative account of the Bible and of humanity's beginnings. I am guessing, but at the very least it's strange that not one of Marcion's books has ever surfaced. Which is a bit funny.

"So, Brad, I reckon that's how we 'forget'. We're told to. If the authorities say this is what happened and you're crazy if you think any different, then remembering takes a lot of courage and can bring you a heap of trouble. It becomes an act of rebellion."

Brad paused for a moment and quietly massaged his right eyebrow to help him absorb my argument. Finally he spoke.

"Planet of the Apes?" he said. *"Planet of the Apes! Clearly your problem is that you're having some trouble distinguishing fiction from history. Planet of the Apes is fiction. There have been, what, seven versions of Planet of the Apes, because it's a great story. That's what*

mythology is. That's what the Sumerian tablets are. That's what the Genesis stories are. That's why you're finding several versions of the same story. Because it's a great story. The fact that something repeats doesn't turn it from a fable into fact. You're reading these mythologies like a fundamentalist. Shouldn't the fantastical content tell you that the stories are nothing more than very old fiction?"

I didn't pause to point out that there are some things he and I would both hold to be true that to others might seem pretty fantastical.

"Fantastical doesn't have to mean zero fact though does it?" I offered. *"For instance, when I went to see "The Madness of King George III" I came home thinking what an amazing story it was and wondering what parts of it might be true. That's a perfectly reasonable question isn't it? So I took myself to the local library and read all the history I could find on the period. Turns out that – other than in the license of screen playing it – it was all true! Same when I watched "Downfall." Fantastical! But it turns out, meticulously researched, word by word!*

"You ask 'How could such a huge part of our memory have been forgotten?' I would say it hasn't been forgotten. Almost every culture around the world has found a way of remembering it. They retell it each in their own traditions and artwork and mythologies. Alright, you can laugh at me and say 'Poor Paul's forgotten how to distinguish fact from fiction!' Perhaps the ancients who authored the world's mythologies would laugh at us and say that we've forgotten how to recognize history when we see it."

Even after a few more beers, Brad and I eventually had to agree to differ. His point was fair when he said that repetition from one mythology to the next was no proof of facticity. And yes, it was only to be expected that Abraham and Sarah would have carried memory and mythology from Sumerian culture into Hebrew memory and culture. In fact, the knowledge of Genesis borrowing from Mesopotamian sources is really nothing new. An array of books on the subject hit the academic bookshelves back in the late 1890s.

So how is it that this same knowledge can still scandalize people today?

Partly I think it's that it doesn't make for very inspiring preaching, and if the preachers don't teach it then most followers of the Bible won't find out. The other part of the equation is that the news cycle moves on. A slew of books comes out on one topic. Next season it's celebrity cookbooks! What's hot on the evening news can be megaphoned one day and buried the next and nobody batts an eyelid.

For many years I had simply accepted the general view of Marcion as an unfortunate maker of mistakes back in the day. Now I had to wonder if Marcion, and the considerable network of churches that fell in line with him, might have been on to something.

Despite their foundational roles in orthodox Christianity, most Christians would have little or no awareness of the ideas put forward by Marcion and Origen. Believers may be familiar with their orthodoxy while completely unfamiliar with what the alternative views were that got ruled out back in the day. There is such a thing as informed orthodoxy. But for the faith community at large, defining and policing mainstream doctrine is itself an exercise of brushing other interpretations away.

There is, of course, a very deliberate kind of forgetting that can happen when cultures have things they may positively wish to forget. For instance, when Britain was at war with Germany a fair bit of name changing was done among our royals to help the British people feel more comfortable with the Saxe-Coburg-Gotha family holding the British Crown. As in the previous war with Germany, members of the royal family whose German accents were too pronounced were kept carefully out of public earshot. After the war the previous enthusiasm of senior royals for deals with the Nazi leadership and alliances with Herr Hitler were tactfully not mentioned. No need. It would be insensitive.

It wouldn't be helpful.

For the people of Israel while in exile, the culture of their Babylonian and Assyrian captors would have evoked exactly the same kind of fear and loathing that Nazi culture later evoked among the allied countries. To the Jewish exiles any fellow feeling, any sense of a cultural debt to their Babylonian or Assyrian captors would have been totally unpalatable. Every cultural and religious difference would have been emphasized to the max and every point of brotherhood and sisterhood diminished. Nothing that felt or smelled even remotely Babylonian would be likely to reach a Jewish sermon of that period.

Now if shared memory is not shared and skips a generation, it is all but forgotten. Israel was exiled in Assyria for a decade and then in Babylon for seventy years. That's more than three generations. And it is during that precise period – the Babylonian exile of the sixth century BC – that many scholars believe the current version of Genesis was redacted – by which I mean changed from an earlier version. Time and reason for a lot of forgetting.

Whatever the faith community we are part of, we preachers always have to be careful not to stray too far from the official storylines, regardless of what longer memory or close attention to the sources might ask of us. The background of group-think, secret shibboleths and unspoken community guidelines all conspire to keep us squarely on the rails.

But when you're holed up in a shipping crate for weeks on end while your traction device works away at your leg, the pressure of group-think is no longer there. Your thoughts don't need to neatly resolve in the few short days before the next sermon is due. You have the freedom to push the boat out a little further and pursue your questions further.

To get any further though, I needed to find a Zaius or two to help me; some guardians or some artefacts of the old, forbidden story. Might there be objects in our own museums and protected areas pointing to a more complex prehistory? Could there be vestiges of our ancient heritage hidden in our very DNA? These were avenues I had to follow.

In *Planet of the Apes,* when General Thade takes the pistol from elder Zaius, he fingers it and sniffs it. He is clearly unhappy that the very existence of this artefact instantly shoots to pieces the apes' erroneous mythology of beginnings – a mythology which framed the world that General Thade lived to protect. How ready was he for his cherished world to be overturned?

The analogy made me wonder. What solid, tangible artefacts might exist in our time, on our planet, that might overturn our own cherished beliefs and suppositions? And what new story of beginnings might emerge from them?

CHAPTER FIVE

KNOWING IN OUR BONES

There was no doubt. They were human. The question wasn't *"what?"* but *"why?"* Why did these ancient Peruvians from more than three millennia ago have skulls that were 60 percent heavier than the average modern human skull? Why did their craniums have only one parietal plate instead of two? Why were their brains 25 percent bigger? And why did their skulls recede so much further back than, probably, yours or mine?

The long skulls I am referring to first surfaced on the southern coast of Peru, on the desert peninsula of Paracas in 1928, discovered by the archaeologist Julio Tello. At first scientists believed the unusual skull shapes might have been the result of local cultural practices which used wooden boards and cloth bindings to reshape babies' skulls. Practices of *"artificial cranial deformation"* were known in South America, as well as further afield in dynastic Egypt, among the Alemani people of Germany, the Alan people of Iran, and the Andean Nazca people.

Closer inspection revealed various clues in the skulls' morphology and volume which indicated the unusual shape was in fact entirely natural. In other words, they were just another type of human being. The largest contingent yet found are the Peruvian long skulls. Others have been found in the Caucasus area in between the Black Sea and the Caspian Sea.

So it is that our Paracan cousins provide us with another anomaly – another clue in another discipline – indicating that the single-line story we have told ourselves about the evolution of human beings is not the whole story.

What's the single-line story? That's the explanation you and I learned in school of the gradual improvement of hominins, beginning with Australopithecus, refining and refining until we reach *"Us"* – *Homo sapiens*. Apparently we just outcompeted every other sub-human genus because we are better, and hey presto – modern civilization. Simple!

One of my school text books described *"our"* arrival on the prehistoric scene. The image on the page, the artist's impression of us, stood out from the depictions of all the previous hominids. We – in the picture – were white and fit. The adjacent text amplified the picture in these words:

"All of a sudden it happened. The group [of Neanderthals] was gathered around the fire, the men were working and talking about the hunt, the women scraping skins, the children playing. Unexpectedly a cry, a hundred cries. And then there appear, as if from nowhere, men of a different type, men who are tall, strong, handsome. Merciless men who spring to the attack. The Neanderthals don't even have the time to comprehend what is happening. A lethal rain of spears pours down upon them. The attackers leap forward to finish off the wounded, to capture and massacre the fugitives. As he fell, pierced by a spear far sharper than his own, he may have thought in his dismay that it was impossible to resist such an enemy…The horrified Neanderthals who had managed to escape took refuge in the mountains where they were in time to perish of cold, hunger and worse still, of loneliness."

"Homo sapiens had developed to a state of perfection. He was now arriving majestically to take his place in the story of humanity." (Man Emerges – Mino Milani – tr May Hope – pub Tom Stacey)

Wow! A state of perfection! Tall, strong and handsome. And white.

Handsome I may be. Tall and white, not so much. So I guess that for me the story didn't quite seem to sit right from the beginning. More importantly, though, my unease primed me to take note over the years of little clues here and there that all was not so neat and tidy in the annals of prehistory.

Back in the 1990s paleontologists discovered the remains of a prehistoric village in France. The skeletons of its last residents revealed the bones of us and Neanderthal people – evidently all living together. In an instant our story changed. Neanderthals were people! They were part of us.

Accordingly, artists in forensic facial reconstruction were given the task of rethinking how our Neanderthal ancestors may have looked. So it was that one Friday morning, as I flicked through the pages of the London Times, I came across a picture of a Neanderthal man, looking considerably less apish than he had when I was a boy. He still had a low forehead with a heavy brow. He still had a heavy jaw and funny, close-sitting hair, and was altogether not quite as tall and handsome as you or me. But he was definitely a person.

This makeover was such a change in the story of the universe that the Neanderthal's image stuck in my mind. I wondered if we might find evidence that Neanderthal people had not all died of cold and loneliness in the mountains after all. Perhaps we had simply all interbred, and their distinctives are still there in the great melting pot of our DNA.

The following week I took a train up to Leicester to visit a friend of mine, studying for his PhD. After lunch on the first day, he challenged me to a game of pool in the college games room. First we had to wait for the university's reigning pool champion to finish his game. He was from Macedonia and a fellow PhD student.

When he stood up and looked at me I couldn't believe it. It was the guy out of the picture in the London Times from the previous Friday. Spitting image. Same forehead, brow, chin, hair, and slight stoop. A PhD student and pool master.

It was a weird moment and I know it doesn't count for much in the great scheme of things, but it blew my prejudices open concerning our Neanderthal heritage and primed me to notice the trickle of paleontological and DNA findings over the next couple of decades underlining the reality of a far more diverse human population in ancient times. Breeding among Neanderthals, Denisovans and Homo Sapiens has now been demonstrated in the DNA record a number of times over.

And then we found Otzi.

Otzi was gradually thawing in a melting glacier in the Italian Tryol. He was 5,000 years old – from a time when, according to my text books, he should have been shivering in a cave, wearing crude bearskins and going *"ugh"*. This was not that. Otzi was wearing well-fitting clothing with a range of stitch-work, different textiles, and insulation for his clothing and shoes. He had shaved. He had fixed his teeth. He had a pouch of carefully crafted utilities. He was us!

Otzi the ice man's sophistication actually reinforced one aspect of what was said in the textbook I quoted from earlier. We arrived *"all of a sudden."* Already clever. Already technologically minded. How do we explain that?

Then again, all of a sudden, somewhere in ancient Sumeria, from out of nowhere, we find a familiar package of phenomena: farming, irrigation, city streetscapes, canals, sanitation, writing, math, banking, astronomy, and time-keeping. And then, all of a sudden these things sprang up all around the world. The artwork and artefacts of these new societies paint a strange picture indeed – a panoply of beings, small and large, human and hybrid, all pointing to a far older and far more complex story of beginnings for the human race.

The more we study human beings, the more interesting and mysterious our beginnings become. What we are now unearthing, and what a new era of DNA testing seems to confirm, matches the folklore of many indigenous cultures – namely that human beings come in all kinds of shapes and sizes, and that in times past the diversity was even greater. Perhaps the hobbits, trolls, hairy people, little people and giants of legend are more empirical than we gave them credit for.

Take giants for instance. Every student of Latin is familiar with the writer Pliny. He was the Governor of Bythinia in the early 100s century AD and kept up a correspondence with

Trajan, the Roman Emperor – a correspondence bequeathed for generations of Latin scholars to pore over from that day to this. His letters include questions to the emperor concerning how Christians should be dealt with. He also refers in passing to a giant of 9'9" [2.97m] in height who was brought to Rome from Arabia where he was feted as a demi-god.

Pliny's near contemporary, the Jewish historian Flavius Josephus, also makes a passing reference to the flesh and bone reality of giant human beings. Josephus is known to many Christian readers on account of his historic reference to Jesus as being the Messiah and his description of the followers of Jesus being defined by their belief in his resurrection from the dead. In his commentary on Jewish history, Josephus makes references to the giants of the Biblical narratives – the Nephilim, Emites, Gittites Anakites and Rephaites. References appear in *Genesis, Deuteronomy, Numbers, Joshua, I Chronicles, II Samuel* and *Amos*. Goliath the Gittite had a height of 9'9" (2.97m) Og the Rephaite, who was king of Bashan was reported as having a bed that measured 13.5'x 6' (4.1mx1.8m)

Josephus traces these people groups back to the references in Genesis 6 to the offspring of Powerful Ones (whom he describes as *"fallen angels")* and human females. Josephus was writing as an historian and so it is interesting to note the equations that he makes in his references.

Josephus also references the following accounts from the Jewish history:

"A giant named Ishbibenob, who was carrying a bronze spear that weighed about three-and-a-half kilograms, and who was wearing a new sword, thought he could kill David. But Abishai, son of Zeruiah, came to David's help, attacked the man and killed him." (II Samuel 21:16-17a)

"Another battle took place at Gath, where there was a giant with six fingers on each hand and six toes on each foot. He was a descendant of the ancient giants. He defied the Israelites and Jonathan, the son

of David's brother, Shammah, killed him. Those...who were killed by David and his men were descendants of the giants at Gath." (I Chronicles 20:5-8)

In his retelling of the story Josephus attributes a height of 8' [2.75m] to the giant victim of David's nephew Jonathan.

Referring to a battle in Joshua's time he says, *"At that time there were still giants whose bodies were so large and whose faces were so different to normal people that they were shocking to look at and struck terror when you heard them. The bones of these men are still on display to this very day."*

What is interesting is that Josephus draws a direct connection between the Greek mythologies of hybrids and those of the Genesis 6 reference. Don't miss the fact that Josephus regarded these mythological references as memories of history rather than as fiction. In his mind he is confirming the fact of the matter by pointing to the skeletal remains *"on display"* at the time of his writing.

Archaeological finds in more recent times have uncovered skeletal remains of a similar height to the giant victim of King David's nephew. The odd detail of a giant with six fingers and toes finds a curious echo in a nineteenth century find in Noble Country, Ohio where the skeletal remains were found of 8' (2.45m) tall humans (or hominids) with six fingers and toes and a jaw where all the teeth, front and back, were molars.

These kinds of anomalous finds have ranged from Ecuador to Malta and are often in places where local folklore holds memories of oversized people from times past.

At the other end of the spectrum, excavations in Indonesia in 2004 brought to life some other human-like beings who had previously been regarded as purely fictional. These were the *"Hobbits"* of Indonesia. The initial find was made by a team of Australian and Indonesian Scientists who were excavating the Ling Bua cave on the island of Flores in Indonesia. The 1m tall skeletal remains were named *Homo Floresiensis* in honor of their location. But the world quickly preferred the name out of J.R.R. Tolkien's tales of the Hobbit. This was in honor of the new-found hominid's small stature and enormous feet!

What was especially surprising about the find was that these little people appear to have been living on Flores as recently as 18,000 years ago.

With finds like these, we have to concede that the prehistory of humanity would appear to be rather more diverse and a lot more complicated than the single line of gradual evolution that you and I may have learned at school. Even today human beings are so diverse that we are not even a single breeding group. Rhesus and non-rhesus blood means that humans fall into at least two breeding groups – and a breeding group is one definition of a species! Looking to our ancestors, we now know of *Homo Sapiens, Homo Neanderthalis, Homo Habilis, Homo Erectus, Homo Denisova, Homo Floriensis, Homo Paracas* and presumably *Homo Nobilis* (that's my name for the molar teethed guys from Noble County). If we include *Homo Gathensis* from I Chronicles 20 then we have people ranging from Low Skulls to Long Skulls, Hobbits to Giants.

Knowing this, one might question the generally accepted interpretation of ancient wall carvings showing people our size sometimes working alongside and sometimes serving much larger people. The often repeated interpretation sees these differences in scale as a bit of artistic license to convey varying degrees of social stature. Yet perhaps what the Bible recalls in print, these wall paintings and carvings are recalling in art – namely that humanity was once an even more diverse population than it is today.

For me the intriguing question surrounding the long skulled people is about why other people groups would go to such drastic lengths to alter the skull shape of their babies to look more like the long skulls? We are naturally and instinctively powerfully protective of our baby's heads. Why would entire people groups around the world deliberately deform their babies' heads to resemble another niche population?

A hint is to be found in the annals of the Spanish Conquistadors in central and South America. They were careful to wipe out the long skulled people as they were recognized as royalty among the indigenous peoples. Curiously, this parallels the images of Egyptian social strata during the period of Ahkentaten, Nefertiti and Tutankhamun. The servant class is represented as people with more familiar shaped skulls. The Pharaohs by contrast have headgear that accentuates or imitates the long Paracas shaped skull. It's an unexplained association that even finds its echo in modern times with the Audrey Hepburn bun bespeaking elegance and sophistication.

We can only conjecture as to why this shape of skull was associated with nobility or social superiority. It could be that the low skull and heavy brow of our Neanderthal cousins were regarded as signs of social inferiority, and that a higher social value was placed on higher foreheads. It might be that with 25 percent larger brains, long skulled people may have enjoyed some intellectual advantages. But given the tiny populations of long-skulled peoples the question remains, who have we all been trying to look like?

When the Paracas skulls hit the limelight again in recent years, some observers wondered if there might be non-human DNA in the long-skulled people to explain both their unusually shaped heads and the social advantage that seemed to come with that head shape. To date the DNA testing confirms our Paracas cousins as entirely human, a mix of Indigenous South American, Western Asia and Eastern Europe. They are another example of ancient human diversity.

Once we begin to see a bigger human family, we are forced to revisit the question of just how ancient are we? As family with *Homo Neanderthalis,* how far back do we go? And therefore how far back does human memory go?

If you look for evidence of agriculture and civilization you might go back to 10,000BC with the advent of farming and cities. If you look for toolmaking and evidence of cultural practices such as art, craft and ritual practices then you might reach 20,000BC. Look for people of our exact design and build and you will reach something like 200,000BC. Some researchers point to evidence of construction in South Africa from that earliest of timeframes. If we prospect for the earliest use of fire then current consensus puts us in Africa, anywhere in the region of 400,000 to 1.4 million years ago – and in the company of *Homo Erectus*. So just how long have we been here and how many human civilizations have come and gone in that time? And what happens if we take that question to the Bible?

If you read the Bible with a fundamentalist hermeneutic then the genealogies will take you back to about 4000BC. Reality, though, may be more complex. Even with a fundamentalist reading, we must take into account that in the pages of Genesis 1–11 we have a collation of scrolls, collected from plural sources. Then, at some point during or after the time of Moses, the scrolls have been redacted to form the single work we know today. However, the separate identity of the scrolls remains so clear that the literary glue can still be seen as we transition from one scroll to the next. So it is easy to see that the different narratives could stand on their own as they originally did.

Viewed that way, it is possible that as well as the two creation accounts of Genesis 1 and 2, and the two flood accounts of Genesis 6–9, we also have two accounts of the extinction of a culture – one caused by the deluge of Genesis 6–9 and one by other means in Genesis 11. Each time humanity is taken back to a more primitive existence. Each time the catastrophes result in human migrations, loss of language and loss of technology.

Put simply, our Genesis scrolls raise the same question as our paleontological explorations, namely, how long have we been here and how far back does human memory go? How many civilizations have come and gone? How many times might our species have interacted with other extra-terrestrial species? How many cataclysms and reboots of history are buried in our collective unconscious?

There is one near extinction level event (ELE) that sits uncontroversially within the timeline of *Homo Sapiens.* Though I was taught about it at school and had drawn pictures of the megafauna rendered extinct by the drama of it, I don't think I had ever realized how close to extinction humanity really came. Nor had I ever read Genesis 1 in the light of it.

I had long known that the last ice age was a challenging chapter for human beings and beasts alike. I knew of the flash-freezing that caused the extinction of saber-toothed tigers, mammoths and woolly rhinos – a process dramatized graphically in the movie *The Day After Tomorrow.*

But just at that moment of fragility, as if the climatic challenges of the ice age weren't enough of a setback, the planet suffered an enormous impact.

Somewhere around 12,800 years ago, just around the time several ancient civilizations appear to have been instantly vacated, the two mile deep ice-shelf that covered North America and much of Northern Europe was impacted by a force so catastrophic that sky fires destroyed cities around the globe, floods submerged entire city-states, an estimated 75 percent of North America's megafauna was obliterated, and humanity itself came within a hair's breadth of disappearing from the face of the Earth.

This was our encounter with the Clovis Comet – an encounter so world shattering that planet Earth would never be the same again. Understanding its aftermath offers us a glimpse of how our ancestors may have interpreted Clovis and all other previous cataclysms and recoveries. Though I had only ever read Genesis as a creation account, understanding the Clovis event made me consider for the first time that what we have in the pages of Genesis 1 may be not a creation account at all, but a story of recovery.

CHAPTER SIX

THE GREAT RE-EMERGENCE

Picture a world shrouded in darkness and engulfed in water. Its surface is peppered by a few scarcely habitable islands, sparsely populated by a handful of species from before the cataclysm.

This world of darkness and water is the canvas for creation as we find it depicted by Genesis, by the Mesopotamian tablets – and also in a Mesoamerican document called Popol Vuh. Popol Vuh (*The People's Book*) is a creation account from out of the ancient Mayan culture of Central America.

Let me describe this almost blank canvas. The darkness is an atmospheric canopy of ash and soot. The waters are the sea levels raised by unimaginable volumes of water released by the melting of an ice shelf two miles thick, which had previously covered the northern half of the planet.

A growing consensus among experts today is that this was the appearance of much of planet Earth following the impact of the Clovis Comet on the Laurentide Ice Sheet above the Great Lakes, just as Earth was emerging from an ice age.

A black sedimentary layer all over North America, South America, Western Europe and the western part of Asia bears dramatic witness to an ensuing conflagration encompassing hundreds of thousands of square miles, followed by catastrophic flooding. The canopy of dust and soot that resulted from the wildfires then blocked out the sunlight over the northern half of the planet.

Coupled with the cessation of the warming Atlantic current due to the invasion of meltwater into the ocean, the combined forces plunged the world, almost overnight, into an ice age even more severe than before – one known as the Younger Dryas cold period. It was a dramatic and super-rapid transition of climate, marked by the kind of flash-freezing that, famously, could freeze a mammoth mid-mouthful.

As the ice sheet melted, it is estimated that sea levels would have risen between 40 and 60 meters submerging any coastal habitations.

In recent years a number of researchers have uncovered evidences of the Clovis Comet ELE and a new consensus is building. In 2007 Richard Firestone, a staff scientist at the Department of Energy's Lawrence Berkeley National Laboratory, put the theory of the Clovis Comet impact squarely on the map when he found micro-sized balls of metals (spherules) and nanosized diamonds in a layer of sediment dating 12,900 years ago at a dozen different archaeological sites in New Mexico. Firestone put these forward as evidence of a massive comet exploding mid-air and impacting various sites, extinguishing the prehistoric Clovis civilization of North America and sending into extinction 75-80 percent of the continent's megafauna. It is estimated that thirty-six species, including mammoths, mastodon, woolly rhinoceros and saber-tooth tiger went extinct through this event.

In 2017 the University of Southern Carolina's Albert Goodyear published a study confirming Richard Firestone's earlier findings. Goodyear had been researching the question of the disappearance of the Clovis civilisation for more than thirty years. Professor Emeritus James Kennett of the University of California is one of a growing body of academics who support the view that it was this comet or asteroid impact that triggered the Younger Dryas period and all its extinctions. The academic establishment can often be slow to embrace new discoveries and perspectives, so the growing acceptance of the Clovis Comet ELE is noteworthy.

It is also intriguing to note what else was happening in other parts of the planet at the same time. A number of city states seem to have been stopped in their tracks. Gobekli Tepe, an ancient feat of civil engineering in Modern Turkey was very carefully filled in with rubble to protect it from something – perhaps an unfolding cataclysm? A maze of subterranean tunnels stretching from Scotland to Turkey testify to a time when people lived underground – presumably to avoid uninhabitable conditions on the planet's surface.

In the seas around Malta, India and Japan the remains of engineered structures appear to predate the timeline of our current civilization. In the Gulf of Cambay on the west coast of India, divers studying pollution levels in the bay discovered the ruins of a city at a depth of 36 meters. Subsequent studies have revealed sandstone walls, a grid of streets and the remains of a seaport. The city, named Dwarka, correlates with a metropolis previously only known in Indian mythologies.

On the island of Malta there are stone vehicular tracks which begin on the land and extend into the sea to depths of around 30 meters. Off the coast of the Japanese island of Yonaguni are pillars, terraces, pathways and guttering at a depth of 36 meters.

All these structures were constructed on land which would have been above water no more recently than 12,000 years ago, prior to the last ice age. All testify to older, forgotten civilizations.

The Clovis Comet is itself named after the civilization, the study of whose remains on the continent of North America, led to the comet's discovery.

These archaeological finds, and others besides, raise the same question. Might it be that between 12,900 and 11,600 years ago the human race had to bounce back after a cataclysm so world-changing that it nearly wiped us off the planet?

What would the memory be of human beings living on the fringe of that catastrophe? Post freezing, fire and flood, famine and re-freezing, what help would those refugee survivors have needed in order to rebuild their lives?

First and foremost, for any kind of normalcy to return, the dark canopy of dust and soot and ash shrouding the light of the sun would need to be cleared. The flood waters would need to be defended against and habitable land reclaimed. The land would need to be rehabilitated for life on the surface, for the cultivation of crops and the husbandry of livestock.

As The Powerful Ones went about these tasks, the surviving human population on these isolated pockets of land would see first the sun, then the moon, and then the stars. Next would be the re-emergence of useful land, then vegetation and animal life. In short, the emergence of the world as we know it would be remembered much as the story goes in the first chapter of Genesis.

Furthermore, a significant clue remains in the Genesis texts as we have them today, that suggest what we are reading is actually the account of a recovery of the planet and a recovery of life. Within our conventional reading this clue would give the appearance of another anomaly. However, once we reframe the story and read our creation texts through the lens of a global relaunch, suddenly this verse makes perfect sense.

The clue in question in the story of Genesis 1 is that before any of the work of *"creation"* has begun the Earth already exists – covered in water and shrouded in darkness.

How can the Earth exist – and water – and even darkness – before light, before the sun or stars? How can there already be an Earth, already covered in ocean, before the first word of creation has been spoken? The Hebrew words describing the state of the planet – *tohu wa bohu* – imply a chaotic, empty wasteland. Is it possible that the text is showing us Earth post-cataclysm, flooded and shrouded in dust and soot – just as in the ELE (extinction level event) that wiped out the dinosaurs?

The picture painted by Genesis 1 matches exactly what we would expect to see if a superior species landed on the planet to help the human species bounce back after a global catastrophe. The primordial step would have to be the separation of salt waters from fresh waters for drinking and agriculture, along with the clearing of the atmosphere to allow the sun to drive all the natural forces that fuel life on Earth. Step two would need to rehabilitate tracts of land for habitation and cultivation. Step three would have to be the replenishing of animal and human species. In short, the recovery sequence would be remembered in exactly the way that Genesis outlines.

The themes of darkness and a separation of waters and the motif of the Earth already existing when the Powerful Ones arrive and begin their work are recurring themes. They are a thread which ties together the Mesopotamian accounts, the Biblical narratives and, as I mentioned a few moments ago, the Mesoamerican mythology of Popol Vuh.

The Popul Vuh document was discovered by a Spanish Dominican Priest by the name of Francisco Ximenez in the early 18th century. It was a tradition maintained by the indigenous priests of the *"Feathered Serpent"* and which Father Francisco translated into Spanish from the Mayan dialect of Quiche.

Popol Vuh is another family's photo album and the experience of flicking through its pages raised my eyebrows even higher than when I first sat down with the texts of the Mesopotamian tablets. The texts of the cuneiform tablets make for percussive and dense reading. The Popol Vuh is astonishingly clear.

In its pages the parallels are clear as the narrator addresses the question of exactly who came to nurture the beginning of humanity and the world around us. The Mesopotamian tablets say it was Superiors who did this. My reading of Genesis says it was The Powerful Ones. And I have argued that those narratives represent the intervention of two extra-terrestrial species. This is the story told by Popol Vuh:

"All was immobility and silence. In the darkness, in the night, only the Creator, the Former, the Dominator, the Feathered-Serpent, those who engineer, those who give being, hovered over the waters as a dawning light."

Once again creation begins in darkness, with entities

"hovering" over the waters. What a fascinating correlation with the language of Genesis 1. And note that they are plural entities; *"those who give being...those who engineer."*

And what do they engineer? First comes light – or dawn. Then humans emerge – or re-emerge. (The word in the text of Popol Vuh can be translated as *"turned up"* or *"showed themselves."*)

While human beings show themselves, the engineers are ***"in the darkness and in the night...holding counsel on the production and growth of trees and creeping vines, of sentient beings and humanity."***

If Genesis and Popol Vuh carry memories of what was in reality a global recovery, somehow it is easier for my small brain to conceive of a superior species assisting with that. For me it is a stretch to imagine multi-dimensional beings with the power or technology to form planets or solar systems. Of course, that speaks nothing to its possibility, only to the limits of my imagination! However, for an extra-terrestrial civilization to visit the Earth and be involved in helping human beings get their footing back on the planet, assisting with flood defenses, cultivation of crops and husbandry of livestock – these are things that we human beings do for one another in the wake of natural disasters. Perhaps the engineers who came and hovered over the floodwaters of Earth, and held counsel about the planet's regeneration arrived in exactly that way – whether it was at the time of the Clovis Comet ELE or a cataclysm even further back in human memory. Could it even be that such interventions may have occurred more than once to aid humanity on its way? Genesis, the Mesopotamian tablets and the Popol Vuh all tell a story of ongoing involvement from the Powerful Ones / Sky People / Engineers to bring humanity into being, finesse it and ensure its survival.

We are so used to thinking of the human race in planetary terms as being the top of the food chain that it is a difficult imaginative leap to conceive of ourselves as mid-level players in a wider cosmic community of intelligences. It is hard for us to imagine human beings as the ones being experimented on – whether nurtured, interfered with or even engineered by another intelligence. However, if you thought the Sumerian story of a Council of Superiors engineering us to slave for them was somewhat far-fetched, just listen to what Popol Vuh has to say. In these verses we eavesdrop on the engineers' conference after they have engineered various animal species on Earth:

"Let us try again. Let us make those who will be our avatars, and those who will bring our food. So the [engineers] determined to make man. Of red earth they molded his flesh…"

The verses of Genesis echo the *"let us"* and molded-from-earth details. The Mesopotamian tablets repeat those same motifs along with the purpose of the humans being to serve the Powerful Ones / Superiors / Engineers and supply them with food.

"But when they had made him they saw that it was not good…Man had been endowed with speech but had no intelligence…Again the [engineers] took counsel. It was now decided to make man…and woman…but the result was not at all satisfactory…They existed and multiplied but had neither heart nor intelligence, nor memory of their [engineers]. They led a useless life and lived like the animals. They were but an attempt at humans."

This human-like population is then attacked, dispersed and nearly extinguished, *"save for a few of their descendants who now live in the forests as little apes."* It is curious that this near extinction is achieved through attacks by animals and by a flood. More curious still is that the Popol Vuh identified the remnants of early versions of humanity with small, ape-like creatures who live in the forest. The creation narrative of Popol Vuh put forward this connection centuries ahead of Charles Darwin and what we know as the story of evolution.

"Once more the [engineers] conferred and the Creator and

Former made four perfect men...They had neither father nor mother. Neither were made by the ordinary agents in the work of creation, but their coming into existence was the result of an extraordinary miracle, a direct intervention of the Creator."

"Now the [engineers] could look upon beings who were worthy of their origin – the Four Progenitors of the human race – strong and good looking. They had clear sight and saw all things – great and small – on earth and in the heavens.

"But this did not please the [engineers.] They had overshot the mark. 'What will we do with humans now?' they said. 'They have become like gods. They have understanding. They will not want to remain subordinate. They will want to make themselves equal to us. Let us therefore limit their sight. So [they] breathed a cloud over the pupil of the humans' eyes...and their sight was darkened."

Another curious coincidence is the interplay in the story between two powerful entities. The CEO in Popol Vuh is named as *"Heart of the Sky."* This corresponds with the *"Air-Lord"* Enlil of the cuneiform accounts. In Popol Vuh the First Officer is named as *"Feathered Serpent" (Quetzalcoatl* or *Totilor* in Mayan versions or *Kukulkan* in the Inca stories.) The Feathered Serpent mirrors Enki, the First Officer among the Council of Superiors in the Sumerian tablets, and his animal association with the Snake of Genesis. In Popol Vuh and the cuneiforms alike it is the First Officer whose initiative upgrades the humans to their final form after a number of experiments.

Greek mythology echoes these same themes in the account of Zeus and Prometheus. Zeus is the CEO who lives in elevated superiority on Mount Olympus. One day he learns that Prometheus – one of the gods ranking below him – has unilaterally upgraded humans' capabilities (specifically in technology.) Like the Serpent in Genesis, Prometheus is considered a wily trickster at odds with his CEO. Like Enki in the Sumerian tablets, Prometheus is associated with wisdom, science and technology. Like Enki and the Feathered Serpent of the Popol Vuh, Prometheus is associated with the hands-on work of upgrading humans from a primitive animal state.

Zeus considers the final upgrade of human beings at the hands of Prometheus as being unwise and potentially threatening to the proper order *of "gods"* and humans. Accordingly, he requires Prometheus to downgrade the human's intellectual ability to reinvent the technologies Prometheus had allowed them. He then punishes Prometheus with eternal torment just to make his point and re-establish his authority.

Once again, in the Greek version of the story, it is the CEO who orders the re-lowering of human capacity and the lower ranking entity who is credited with the (now lost) upgrade. It is an odd motif to find repeating from culture to culture. The repetition cannot be coincidence. But how are these accounts connected?

In Popol Vuh the First Officer so involved with the development and upgrading of human beings is depicted as a feathered serpent. So it is intriguing to find beings with large, strong, human-like bodies, with scaly skin, feathers and bird like heads figuring repeatedly in the stone carvings and written mythologies of many cultures around the world. Berossus' Greek account of Babylonian mythology depicts scaly Oannes and the Apkallu. Similar figures appear in the lore of Bolivia, Ecuador, India, Tibet, Japan, Thailand, Indonesia and in the art and literature of the Mayans, Aztecs and Olmecs of ancient Mexico, all carrying the same mysterious bucket or handbag.

In the case of the Snake-Power of Genesis, Enki of the cuneiforms and the Feathered Serpent of Popol Vuh, all are involved in progressing the human experiment and specifically in upgrading the intellectual capacities of human beings. Perhaps the most significant pattern that repeats across Genesis, the tablets and Popol Vuh is that the emergence of modern humans is portrayed as the outcome of a series of experiments resulting in a gradual, staged development.

Popol Vuh describes the creation of people from white and yellow corn. The Mayans were the people of maize.

In the past people read this motif as pure metaphor, nothing more than an artistic flourish reflecting the veneration in which the Mayan's held Maize, honoring its importance to their existence as a people. Recent research adds another layer of importance to this agricultural connection. Without doubt the development of maize as a farmed grain was critical to the establishment of Mayan civilization. But the picture now emerging is of a deeper and far wider significance.

Twenty years ago some research was undertaken by scientists at the Agricultural University of Norway in As, Norway, and the Max Planck Institute in Cologne, Germany. They found an area in the Fertile Crescent near the Karaca Dag Mountains, in Southeastern Turkey, where a whole number of cultivated foods all made their first appearance – and within a very short time of each other. These foods included wheat, barley, peas, lentils, broad beans, chickpeas, grapes, olives, flax and bitter vetch. Close by was evidence of the earliest domestication of sheep, pigs, goats and cattle.

Bruce D. Smith, director of the Archaeobiology Program at the Smithsonian's National Museum of Natural History, said of the discovery, *"It is a remarkable congruence of plants and animals all domesticated in a relatively small geographic area very early–a remarkable jackpot."*

No one knows who first cultivated these plants. The researchers speculate that it may have been a single community, perhaps a single family, who first stumbled on the idea of agriculture. The team's analysis of the parent plants' DNA found that it took a modification in just one or two genes to transform the ancient wild wheat into a useful crop. And that was enough to start the first regular cultivation of plants. All these innovations appear to have occurred in this part of the world at about the same time

– just as the last ice age was drawing to a close.

Farming certainly sprang up quickly in other places, too, and not long after the changes in Southeastern Turkey. But the sudden appearance and tight localization of the change in these particular plants points to an incredible sequence of technical advance which is hard to explain. The leader of the research team, Manfred Heun, at the University of Norway's Department of Biotechnological Sciences, spoke about the findings to *Science* magazine in 1997 and said, *"I cannot prove it, but it is a possibility that one tribe or one family had the idea."*

Maybe. Or did we have help? Look at that timing. This sudden progress in the cultivation of crops, allowing the development of cities and advanced civilization occurred just after the last ice age, in that critical recovery period when things could so easily have gone another way and when humanity desperately needed a helping hand to get going again.

Before we get too sucked into the moment, we need to keep in mind that this incredible leap forward in agriculture was the beginning of our civilization – not of all civilization. Under the seas of India and Japan, buried in the rubble of Turkey at Gobekli Tepe, and hidden in the canyons of the Clovis people's territory in North America lies the evidence of previous technological civilizations arrested by some natural disaster long before the community at Karaca Dag began to cultivate the foods we still eat today.

It is disturbing for us to acknowledge that advanced civilizations like our own can be extinguished. The last ice age reminds us that climate change can do it. The Clovis Comet ELE reminds us that cosmic factors such as asteroids or solar flares are enough to do it. In the twentieth century our current civilization learned how technology can do it.

All of these may have happened in our long forgotten past. The book of Ecclesiastes says,

"What has happened before will happen again. What has been

done before will be done again. There is nothing new in the whole world. "Look," they say, "here is something new!" But no, it has all happened before, long before we were born. No one remembers what has happened in the past, and no one in days to come will remember what happens between now and then." (Ecclesiastes 1:9–11)

If we really believed that past civilizations have been snuffed out just like that, I wonder if we as a civilization would live a little differently? If we believed that our civilization is as vulnerable as previous ones to random impacts from interstellar objects, would the technological imaginings of movies like *Deep Impact* and *Armageddon* need to become technological fact. Would developing our spacefaring capabilities become a higher priority?

The famous astrophysicist Carl Sagan wrote, *"Since, in the long run, every planetary civilization will be endangered by impacts from space, every surviving civilization is obliged to become spacefaring– not because of exploratory or romantic zeal, but for the most practical reason imaginable: staying alive... If our long-term survival is at stake, we have a basic responsibility to our species to venture to other worlds."*

What if we believed there was nothing we could do about the fragility of our existence? Would we re-prioritize the sharing and development of new technologies to get us off planet? Would we give more thought to our quality of life as a civilization? Would we invest more into our way of being and our relationship with the eternal?

Without doubt, ours is a very vulnerable civilization. You and I are daily dependent on all kinds of technology that we know how to use but which neither of us can make. Put each one of us into a post-apocalyptic world, and we would quickly find out how many basic technologies we would need help in recovering.

Today archaeological sites and artefacts demonstrating older, forgotten civilizations are painting a picture in which the creation of agriculture in Karaca Dag is not a story of origins but a story of recovery – a re-emergence of *Homo sapiens* from near extinction. If ever there was a time that humanity needed a helping hand from interstellar neighbors, it would surely have been that moment.

The testimony of the Sumerian and Babylonian mythologies is that it was the helping hand of the Sky People that modified the crops and gave human society the technical know-how to cultivate them. In the Babylonian tablets, a female from among the Sky People teaches the humans to cultivate plants and gives them bread and beer – two foods that result from cultivated grains.

The Popol Vuh speaks of the creation of the people who are made from maize. The Mayans were the maize people. Another echo resounds in the creation story of the Zulu people, which praises a female god Mbab Mwana Waresa for helping the first people to establish themselves. She does this by teaching the first community of people how to farm and how to make beer.

The coincidences begin to pile up.

The sudden genetic modification of grain at the same time as the sudden appearance of early civilizations poses an intriguing chicken and egg kind of question. And that this should happen in the crucial time of humanity's great re-emergence, post-cataclysm, was a discovery that caught my eye. It made me wonder about the longevity of memory and how it resurfaces in our story-telling from generation to generation, from culture to culture; from mythology to fiction to film-lore. It made me wonder if our ancient creation myths are in fact the scars of ancient traumas – a strange fusion of what we remember and what we have forgotten. And it made me wonder if even the most florid metaphorical-sounding language of narratives like Popol Vuh might contain a greater depth of memory than we have ever imagined.

CHAPTER SEVEN

THE MEMORY OF US

It is an amazing privilege to sit at the feet of the elders and story-tellers of our ancestors. We might respectfully hesitate before wading in as if we totally understand the territory being described to us. On the other hand, we only impoverish ourselves by ignoring or remaining agnostic about what we have been shown.

Until recently, the mainstream tendency has been to regard mythology as a mélange of fiction and moral tale. In faith communities there is often a pull in the opposite direction, towards a fundamentalist interpretation which reads the texts as literally as possible. Every preacher knows the subtle dance shaped by the presence of these polar opposites in every congregation.

As I continued my journey through Genesis, Popol Vuh, the cuneiform tablets, and other accounts besides, noting all the parallels and correlations, my sense was growing that even the most metaphorical of myths may turn out to be the vehicles of ancient memory.

In a plural-elohim reading of Genesis, in the cuneiforms of Mesopotamia and the Popol Vuh of Mesoamerica, among the themes that repeat is the assertion that modern human beings emerged through interaction with another intelligence, and that our evolution was something that happened artificially and in stages.

Popol Vuh's account matches the Sumerian story, beginning with the development of male avatars or slaves for the superior beings. Only in a later stage is humanity modified to become a fully-fledged, fertile, male and female species all its own. In Popol Vuh, just as in the Mesopotamian tablets and in Genesis, we humans ultimately become a species whose progress creates angst among the superior beings. In Popol Vuh, the engineers convene an urgent conference to discuss the problem. As we listen in, we hear them closely paraphrase the anomalous verse from the book of Genesis, *"Now they have become like one of us!"*

If that correlation weren't enough of a waving flag, the ancient elders of the Efik people of Nigeria would tell us that in the beginning their gods had the exact same conversation. The Efik tell the story of the first human couple formed by two powerful creative Sky Beings – Abassi and Atai.

In the beginning human beings were like children. Their understanding and intellect were simple and innocent. Unable even to feed themselves, they need the Sky-Being Abassi and his wife Atai to supply their every need. And at first the human beings are content to be looked after this way.

However, after a season there comes a day when the humans tire of living in the home of the Sky Beings. They feel that they are ready to live on the Earth. Abassi, however, has qualms about this next step. He sees the Earth as a place where the humans' knowledge and understanding will grow and mature. He fears that the human beings will become too developed and come to match his own level of wisdom. Anxiously, he confers with his wife. He tells her that he does not want the humans to become *"like one of us!"*

Atai ponders the matter and determines a compromise. The humans are to live on the Earth by day but must return each day to their home with the Sky Beings, to eat and rest.

At this stage, the first humans are forbidden to marry or have children, or hunt or farm. This limit has been set in order to prevent the humans from producing a nation strong enough to challenge the power of the Sky Beings.

But after a season, the woman tires of being treated like a child. So one day, she simply refuses to return with the man to the home of Abassi and Atai.

The next time the man sees the woman, he finds her working in the fields. She has learned to farm and is growing her own food. The man is impressed with her wisdom and independence and quickly decides to join her and help her. The two soon fall in love and never return to the sky-base again.

After many years, the humans become a great people, living on the Earth and working in the fields.

One day Abassi goes down to the Earth and into the fields, where he is horrified by what he finds. The humans have grown significantly – both in intelligence and in number. Abassi's fears have been proven.

When he returns home to the sky, Abassi shares his concerns with Atai. Atai now produces an emergency plan. In order to combat human development she sends death and suffering into the world and causes the humans to be in perpetual conflict. The first man and woman die immediately. And their descendants? They have experienced conflict and suffering on Earth ever since.

The number of correlations between the Efik account and the three mythologies we have already compared really is astonishing.

The ancient Efik elders are singing us a song with many familiar notes. Their story tells of a staged development of human beings, engineered by powerful Sky Beings. These beings begin the experiment with non-self-aware, non-sexualized beings, who have to be corralled and cared for. Another correlation is in the fact that food, self-awareness and intellectual progress are offered to the man by the woman – defying the wishes of their creators. The initiative results in childbearing and punishment. Parallel after parallel.

The strange motif of the threat of human numbers echoes from the Sumerian account and Atai's cruel response in the Efik story echoes the limiting of human lives, the mass killing of the great flood and the fragmentation of Babel.

All four creation accounts repeat the theme of creators pitching themselves against human progress. It is a strange motif and it is hard to imagine in whose interest such a story would have been created. It would appear to glorify no one in the story.

In the Mayan story of Popol Vuh, the engineers share the same anxiety about human progress and intervene decisively to limit the humans' capacity to cause their masters too much trouble! The engineers achieve this with a downgrade of the humans' vision and understanding. The engineers' third experiment in engineering people resulted in human beings who were capable of seeing beyond the limits of earthly, physical reality. To make humans more manageable the engineers take this higher faculty and switch it off. It is the final tweaking of the human condition as described in Popol Vuh.

There is a disturbing parallel in the Genesis story of Babel. In that account, human society has recovered after a great cataclysm – the flood – and rebuilt itself to the point that a city has been developed in Mesopotamia, in the region of Sumeria. The city would include a tower with a stargate. Clearly alarmed by this last technological detail, The Powerful Ones confer in Council as to what should be done to manage the advancing capabilities of the human race.

"If they are capable of this," they say, *"nothing will be impossible for them."*

Curiously it was only after I read about the switching off of our higher faculties by the engineers in the Mayan story that the profound cruelty of the Babel event in Genesis really struck me. When we read Biblical narratives as God-stories we are programmed to ignore or excuse quite monstrous actions – because they are understood to be the actions of a holy God. Yet surely His ways are higher than ours not lower! As soon as we reframe the stories as the actions of Powerful Ones the morality of the actions can immediately be seen with shocking clarity. Suddenly genocide is nothing other than genocide; an act of retribution is nothing other than an act of retribution.

The current version of the Babel story in Genesis reports it as a story of Yahweh. However the plural forms, the language of conferring, the action of coming down and the *"let us go down and confuse..."* all clue us that this is in fact another *elohim* story – as indeed it has to be, being placed in time centuries ahead of Yahweh's appearance to Moses. Our redactor friend J has done little more, it would seem, than insert the holy name into an *elohim* story.

According to our mainstream Biblical translations, the divine punishment that is meted out upon the humans is God's retribution for the terrible crime of infringing divine zoning laws, constructing a building that is too tall, and his punishment is for the arrogance of trying to reach the heavens. Read that way, the Babel story would surely be a bizarre overreaction on the part of God. Not to mention that we have erected taller buildings in the ages since without any outpouring of divine wrath. Something is clearly off-key in that picture.

In fact, even in our conventional translations there is a clue apparent that all is not as it seems. *"What exactly does 'reaching the heavens' involve?"*

As I suggested earlier, a tighter translation of the word *babel* indicates that the tower is intended to house a stargate that will literally enable people to *"reach the heavens."* The nuance is confirmed when cross-referenced with the cuneiform telling of the story, in which three hundred observers are dispatched from Babel to their stations among the stars.

The texts tell us that it is the technological ability being built into Babel that the council of Powerful Ones wishes to sabotage. They do not want to see human beings operating on that level. They do not want to be jostling up against a spacefaring human race and so they resolve in Council to shut down the entire civilization.

In the drama of Babel, we watch as a highly developed,

technological society is utterly destroyed, and its city abandoned as the Powerful Ones intervene to stall human progress. In a callous and violent act, the Powerful Ones pull the plug on human civilization once again. This time they do it by extracting from the human beings the faculty of spoken language. As if by inducing a collective stroke the Powerful Ones turn our brains' default settings all the way down, right back to the beginning, to the point of extinguishing our human capacity for intelligible speech. In an instant we find ourselves returned to an almost animal state.

All around the world we find the cities of past civilizations, abandoned and forgotten. We are left to wonder what caused their civilization to falter. How were they not able to continue? Was it disease, flood, fire, famine, climate change or nuclear fallout? In the case of Babel in Mesopotamia, Genesis supplies an answer. The population is incapacitated by an assault from a more advanced and non-benevolent species.

It was an act of unspeakable violence – literally. In Babel we are witnessing the neurological equivalent of bombing human civilization further back than the Stone Age. In that moment, the continuity of the Bible's timeline is lost. What sharing of knowledge, what record-keeping, law, literature or technology would be possible from that point on without shared language?

Indeed, the Bible's stories of beginnings stop right there. Nothing further happens until, in a completely different age, Abraham and Sarah emerge from out of Sumeria. They are children of the ancient culture credited with inventing – or rather reinventing reading, writing, dating, record keeping, agronomy, technology, law and the city. Cuneiform script was created as a method of transcribing the diversity of languages now present in the world. All this from out of the ashes of that same plain of Shinar where human progress had been decimated in the world of long ago.

Abraham and Sarah's journey, beginning in Sumeria, plays

out in the world as we know it – albeit the Powerful Ones are still present. The world before the time of our father and mother in faith requires us to unpick everything we thought we knew about planet Earth, human beings and God. We have a lot of remembering to do.

And if ever we needed an indication that the Powerful Ones / Sky People / Engineers of that *"world of long ago"* were not merely the literary projections of a God of love, we only need to put the incredible anti-human violence of Babel next to the violence of the Flood and the violence of the Fall and we have more than enough of an indication. All were attacks aimed directly against the progress of human society.

We need to rescue our understanding of the True God from being confused with the not nice stories of the Powerful Ones, just as Joshua calls us to *"Forget The Powerful Ones whom your ancestors served in Sumeria and in the time before Abraham and Sarah. Serve Yahweh and know that he is God!"* If we confuse the truth of God with the stories of the Powerful Ones, we end up associating our God with their monstrous acts.

The actions of the Powerful Ones / Sky People / Engineers reveal them as beings of many layers – some benign, some less so. We see them treating human beings much as we would treat livestock on a farm, or in the way human masters might treat their slaves.

Such comparisons don't leave a very pleasant taste in the mouth. Our mythologies do not make for sweet fables that we can tell our children to encourage them to be good and devout. The storylines themselves force us to listen to these mythologies differently and consider what fragments of understanding we are being offered by the elders and memory keepers of ancient times.

We cannot read stories that share so many common motifs as if they were purely creative allegories, which just happen to parallel each other from culture to culture out of pure coincidence. On the other hand, a fundamentalist reading of our mythologies can only survive by ignoring all the anomalies.

Simply reading these mythologies over the months of my forced seclusion was sufficient to break the spell of the fable *vs* fundamentalist dialectic with which we so often approach our stories of beginnings. As I joined the dots from one mythology to the next I could see that the very strangeness of the stories, and the unlikely repetition of those strange motifs stand as evidence that in these mythologies lies a body of ancient collective memory. I might call it collective trauma, because it appears to be the most scarring aspects of our original journey into being that our mythologies recall. Memory has a way of doing that.

If these stories are not the repositories of ancient memory, then you really have to ask what other purpose they could possibly serve. Why would human beings invent such demeaning stories about themselves? These explanations glorify no one. Such accounts of our engineering don't glorify the True God. Neither do they glorify the Powerful Ones, Sky People, Engineers nor any human elites. Neither do they provide an inspirational or motivating explanation for the presence of the human race. Ultimately, there is really no good reason for anyone to invent this as their cultural mythology.

Each indigenous people has taken great care, in its culture, of sacred storytelling. I believe that this lore holds for posterity, for generation after generation, for any with an ear to hear. It is the memory of us – who we are and where we have come from.

If we can embrace this possibility it sends us back to review the long and complex timeline of human development and ask, *"When did the engineers do their work?"* How can we know? And is there any tangible evidence of a higher intelligence episodically interacting with human history?

This was the question I found myself mulling over – but not in my shipping crate. I was in a land far away, standing in front of a twelve-foot tall anomaly in the Indian capital of Delhi.

CHAPTER EIGHT

THE EVIDENCE OF THINGS

I had never experienced a wall of noise like this. Walking into the terminal building at Delhi International Airport in 1976 was an assault to the senses, and, for a diminutive boy visiting from England, a little scary. We were on the Indian subcontinent as guests of the owner of Air India, a connection which helped hurry our passage through the teeming terminal floor and out into a waiting car.

We had many reasons to be excited to be on Indian soil. It is a country of astonishing beauty, diversity, surprises and challenges. As we travelled around the cities and towns of India and Kashmir, the differences of the physical environment, the built environment, the gastronomic and social environment were stimulating and fascinating.

Every temple had towers adorned at their summits by renderings of the vimanas – the ancient flying machines of the Vedic scriptures. They were just a part of the layers of message and memory encoded in India's historic buildings.

I had seen the iron pillar before, 6.5 metric tonnes of extremely high-grade wrought-iron, celebrated by the locals as a phenomenal exhibition of ancient Indian technology. This is because though constructed sometime in the 4/5th century AD, the Mehrauli Pillar in New Delhi has shown scarcely a hint of decay or rust – such is the quality of the iron of which it consists. On that day in 1976 the Qutub Minar complex, in which the pillar now stands, was quiet enough that we were able to touch it and feel it and see if we could wrap our arms around it from behind – a local custom said to bring good luck!

The state of preservation was indeed remarkable. It had stood in the Qutub Miniar complex for the best part of one and a half millennia. But just lately the Mehrauli Pillar had begun to attract more traffic. Something had increased the flow of pilgrims to it from all around the world. The something was the pillar's appearance in a blockbuster of a book by Swiss researcher Erich Von Daniken. And that is where I had seen it before.

Von Daniken's seminal book *Chariots of the Gods* was the progenitor of a whole genre of literature surrounding ancient mystery, ancient aliens, crypto-paleontology and ancient civilizations, all cataloguing artefacts and findings that presented as anomalies to our conventional histories. The credibility of the evidences one would have to place on a spectrum. The implications of some data were inescapably challenging to the status quo of our history. Perhaps some of Von Daniken's data would be examples more of pure faith than of science. And as the letter to the Hebrews points out, *"faith is the evidence of things hoped for."*

We returned to our hotel from our excursion to the Mehrauli pillar wondering if it was an item more of faith than of evidence.

I first encountered Erich Von Daniken's thesis in *The Chariots of the Gods* at a dinner party hosted by my parents. It was always a treat for me and my brother to be admitted to these grown up occasions. My mum and dad always pulled out the most exciting and experimental of dishes and fostered fun and stimulating, grown-up conversation with our guests. Though the book had been in print for eight years it was only just beginning its life as a stimulating and polarizing conversation starter. Which is what I recall from that dinner party of my young youth.

Having bounced the idea around the dinner table for some minutes, my dad concluded, *"I can easily believe that I am the product of a higher intelligence and I can well imagine that one day the highly intelligent species will return and recognize me as one of their own and take me home with them."*

He was joking, of course, and we all laughed.

Yet the ideas of a bigger interstellar community intrigued

me. By that young age I already held the conventions of Christian religion in some suspicion. It was a suspicion that had been inadvertently nurtured by my primary school in Buckinghamshire UK. I loved my days in that school. It was a warm and nurturing environment, just getting going in the new philosophy of education, child-centered learning. Nevertheless, there remained vestiges of a long established, more disciplined order of things. It was in that context that the Headmistress introduced a new hymn to us at the morning assembly.

"Mrs. Clarke's class..." (Mrs. Clarke was the one teacher who could adequately play the piano) *"...will now teach us a new hymn. Please listen quietly, children."*

The hymn began with the words, *"When Jesus was a little boy..."* and went on to describe how, when Jesus was a little boy, he was good and obedient to his parents, never misbehaved or gave any problem of any kind to his teachers at school. So children you had better do the same!

I was only five years old but I still recall my inward reaction. The transparency of the school's use of religion to lick us infants into shape was not lost on me. *"They're just trying to get us to toe the line!"* I said to myself, quietly concluding that this religion business was clearly intended for weak-minded people who are content to be manipulated!

So, when, as a middle school student, I heard that the very idea of God was probably based on humanity's contact with technologically superior species in the ancient past I was delighted to have an alternative view to explore. My enthusiasm for this perspective propelled me into some spirited discussions with a group of born again Christians at High School, who patiently suffered my unremitting deconstruction of their logic over the years that followed. But I was like a moth to flame. I couldn't leave them alone. I wanted them to admit that I had some evidence on my side and that all they had was blind faith. At the same time, I felt in my blood that these Christians had something I could not touch.

However, through these lengthy dialogues I gradually came to understand that the use of religion by Primary Schools and the credibility of Jesus himself are two completely different things. As I read the New Testament to help me martial better arguments against the Christians, I began to perceive that Jesus himself compelled a different response. The Jesus of the Gospels appeared to be about liberating people from hierarchies and from living in fear. That is how my journey as a Christian began.

Meanwhile, proper academics were going to town to debunk Erich Von Daniken.

His book had made a great deal of the famous lines at Nazca, a curious grid of enormous clean lines in the rocky desert mountains of Peru, visible only from high altitude. How could primitive Stone Age man have made them and why?

Now I watched as debunkers reported that the lines are really not that large, that they can be produced by natural processes. Why you could even mark the ground simply by scraping your shoe on a little patch of the desert floor, scattering the darker surface stones and revealing the lighter desert floor. And the locals' use of these lines in religious and cultural rituals was well known. So there was another curiosity suddenly less curious.

Professional academics mocked Von Daniken as an amateur and over the course of time I was able to hear the opinions of professors and academics who knew a little more on the topics into which *Chariots of the Gods* was trespassing.

There was no shortage of mistakes in Von Daniken's writings for the real experts to shine some light on. What was mocked more than anything else was Erich Von Daniken's hypothesis that in the ancient past extraterrestrial species had seeded our part of the Milky Way with spores of their own DNA. Their purpose was to foster life and intelligence similar to theirs on any welcoming planetary environment. In due time they would be ready to nurture and assist the life and intelligence that would result from the diaspora of their genetic code.

Von Daniken pointed out that many indigenous mythologies tell of beings coming from the sky and having sex with our earthly ancestors – people who were themselves the outcome of a prehistoric ET seeding of our planet. This hybridization resulted in the species we now call *Homo sapiens.* These were ideas to be found in the apocryphal Hebrew book of Enoch, the Sumerian tablets, in the Hindu Scriptures and the Greek legends.

But Erich Von Daniken was an amateur. What did real scientists have to say?

One day I heard the great astrophysicist Carl Sagan staking his claim. Von Daniken's idea that we *Homo sapiens* were seeded by ETs was ridiculous. The idea of another species mating with humans was as likely a proposition as a human being successfully mating with a petunia!

So I thought little more of this theory of an ET diaspora for the next four decades, right up until an accident forced me into months of traction, slowing me down to the point where I had time to reflect on questions which had sat latent for decades.

As I read the various translations of the cuneiform tablets during the months of my convalescence a long-neglected door was being knocked upon. The repetition from a plural Genesis to the cuneiform tablets to the stories of Popol Vuh and on and on begged questions of my old conclusions. What would Carl Sagan have to say?

So I took some time to reacquaint myself with a real scientist, one who had nurtured my sense of wonder and enquiry all those years ago. What I found turned my comfortable old conclusions on their head.

The Carl Sagan I knew through the media was the ultimate sceptic in the best scientific tradition. He always appeared meticulous in his commitment to "knowing" as little as possible while provisionally accepting and championing those things apparently confirmed by solidly scientific approaches. Knowing that Carl Sagan was within a hair's breadth of being an atheist, I was astonished when I sat through the movie of his book *Contact*.

This is a movie worth watching. The story dramatizes a subtle play of evidence, belief, subjective experience and intellect. In so doing it reflects profoundly the layers of esoteric knowledge, evangelistic urgency and claims to objective evidence that are wrapped up in the Christian faith. Inwardly, I raised an eyebrow.

It is more than twenty years since the movie's release so, with the caveat of *"spoiler alert,"* I think it is fair for me to tell you that the story involves a multi-dimensional, extraterrestrial intelligence making contact with Earth to encourage humanity in its ongoing development. I raised even more of an eyebrow as I watched these themes unfold in the screenplay.

Carl Sagan wrote *Contact* in 1985 and it hit the screens in 1997. It is his only novel. I began to wonder why this territory lay so close to his heart. As I mentioned before, sometimes what we cannot sell or support as fact we tell as story. I wondered if Sagan's compelling storytelling was speaking for another layer of our human being – a creative layer that intuits and hypothesizes and leans hopefully towards conclusions in anticipation of clinching proof.

What is a fact is that Carl Sagan was openly committed to the idea of ET contact through his public role in establishing the SETI institute, an organization devoted to prospecting for signs of intelligent life orbiting distant stars. After *Contact* I had to wonder if that was the public expression of a deeper, resolutely unspoken faith.

At least I thought it was unspoken. Because in 2017 there came a little flurry of controversy surrounding an analysis of Carl Sagan's work, written by Donald Zygutis. The discussion shone some fresh light on what appeared to be a passing moment in Sagan's literary career.

In 1962, while working at Berkeley, Carl Sagan published a paper called *Direct Contact Among Galactic Civilizations by*

Relativistic Interstellar Spaceflight. In the paper he speculated as to the probability of interspecies contact in Earth's distant past. He wrote, *"There is the statistical likelihood that Earth was visited by an advanced extraterrestrial civilization at least once during historical times."*

On page 497 of his study the *"at least once"* balloons to a possible 10,000 times!

This may seem a bold claim indeed but it was based uncontroversially on the Drake Equation. On the basis of today's data the Drake Equation was a very conservative formula for calculating the probability of life on other planets. However, Sagan added a significant personal gloss when he speculated that the account of Oannes in the mythology recorded by the Greek Babylonian priest was in reality the recollection of close contact with an ET.

On page 496 of his paper Sagan wrote: *"There are other legends which more nearly satisfy the foregoing contact criteria, and which deserve serious study in the present context. As one example, we may mention the Babylonian account of the [generation of the] Sumerian civilization by the Apkallu, representatives of an advanced, nonhuman and possibly extra-terrestrial society."*

Page 29 of Sagan's paper is bolder still. Calling on the Berossus account he speculates the ET Apkallu may have been directly responsible for gifting the ancient Sumerians with the tools of civilization.

Four years later Carl Sagan was prepared to go even further. In 1966 he co-authored a book with Ukrainian scientist I.S.Shklovskii entitled Intelligent Life in the Universe. In it he writes: *"Stories like the Oannes legend, and representations especially of the earliest civilizations on Earth, deserve much more critical studies than have been performed heretofore, with the possibility of direct contact with an extraterrestrial civilization as one of many possible alternative explanations."*

So the Babylonian and Sumerian texts had Carl Sagan's

attention too! At that point he was willing to believe that ancient mythologies, however fantastical and kaleidoscopic, might in fact be the vehicles of ancient memory – specifically ancient memory of extraterrestrial contact.

Erich Von Daniken quoted Carl Sagan's paper and book to undergird his own arguments. But by the time *Chariots of the Gods* was charting, along with all the critical panning and debunking that came with its success, Carl Sagan was steering a different course. This may have been because his convictions had changed and his rigor as a scientist and sceptic had tightened. But in the light of *Contact* I had to wonder what other factors might have been in play.

It is logical to hold the expertise of tenured professors in higher regard than the research of enthusiastic *"amateurs"*. However academic institutions are often dependent on sources biased towards the status quo. In that regard the jibe of *"amateur"* may sometimes be a bit of a cheap shot from those who have the funding, aimed at those who don't. Yet, notwithstanding those conservative energies, there is no shortage of non-amateurs, top academics have been willing to back Von Daniken's hypothesis that Earth had been seeded by ETs. This thesis has been supported and argued for variously by scientists such as Cambridge Astronomer Professor Sir Fred Hoyle, Francis Crick who won the Nobel prize for his co-discovery of DNA, Leslie Orgel, a British research Chemist who with Francis Crick co-authored a paper arguing for *"panspermia"* – the theory of ET seeding – and Hungarian astrophysicist George Marx. Weightier supporters you could hardly wish to find.

As you can imagine, learning all this left a significant dent in my faith in the debunkers. Or to put it another way, I was now beginning to doubt my doubts! All in all I felt I owed Erich Von Daniken a second hearing.

This was the reason I now found myself shuffling through aerial photos of Peruvian mountainscapes. As I scrolled through,

I came across an image I had never seen before and which left me open-mouthed. The image was of an arid mountain range marked with the usual chiseled features and folds typical of any mountain range – except for something inexplicable. The tops of the mountains were sheared clean off. It was as if some giant blade had sliced through the mountain peaks like butter. That is Nazca.

Surely no natural process could do that! No amount of shoe-shuffling or stone-kicking would ever produce a geological feature like that. That's for sure! The incredible improbability of the feature is visible only from high altitude – just as Erich Von Daniken had written. Why had I never seen this? Why in the debunking shows was the scale of this geological anomaly never shown?

I was now eager to give Erich a fairer hearing. Certainly, I could see that not all of the claims of *Chariots of the Gods* stand up to scrutiny. More than once Von Daniken has had to concede, *"I was wrong!"* regarding various evidences. The book was clearly midwifed into being with an enormous burst of enthusiasm and under-zealous fact-checking. Consequently, not all the facts quoted by the book are facts. Not all the book's anomalies are as anomalous as Von Daniken may have thought.

And that may have been the case with regard to the Mehrauli Iron Pillar. In an ironic twist, the accelerated flow of tourists compelled by *Chariots of the Gods* to touch and feel the rust-free pillar had begun to affect the pillar's iron surface, resulting in a layer of corrosion. The pillar had survived virtually unchanged for one and a half millennia but was now being damaged indirectly as a result of Erich Von Daniken's book! Twenty-one years after my visit, authorities placed a barrier to protect the iron – which is in fact of exceptionally high grade – from the sweaty palms of enthusiastic visitors!

Notwithstanding these proofs of Von Daniken's amateur status, there is life in the hypotheses and enough real data in

Von Daniken's catalogue of findings to keep the questions live. In the fifty plus years since the book's publication, a body of significant discoveries unearthed by Von Daniken's successors have only added to the list of archaeological anomalies that point to the possibility of historic contact.

For me one of Von Daniken's most intriguing anomalies was an ancient carving. It was unearthed in 1949 in the bowels of an abandoned city by the Usamacinta River, in the Mexican state of Chiapas. There it lay buried deep under the foundations of an ancient Mayan pyramid.

CHAPTER NINE

WE DIDN'T SEE THAT AND NEITHER DID YOU!

It took three years to empty the shaft of the rocks and stones that filled it. Month by month more of the descending staircase was uncovered. Archaeologist Professor Alberto Ruz Lhuillier began the excavation in 1949 when he discovered a groove hidden in a pyramid at the abandoned Mayan city of Palenque. The groove led to a shaft, descending below the foundations of a stepped pyramid called the Temple of Inscriptions.

In the fourth season of digging, the professor's team reached a triangular stone door. Once the team had broken through the doorway Ruz Lhuillier pushed a torch and then his head through the opening they had made. In a state of shock he called back to his team, *"I don't believe what I am seeing! It looks like a chapel with candles hanging from the ceiling!"*

The candles were stalactites, indicating the antiquity of the chamber, which was about seven meters in length. Further excavation revealed that the chamber was a crypt, the secret resting place of a magnificent stone sarcophagus, topped with a stone slab 3.8mx2.2m.

Carved onto the monolith was a beautiful illustration of a young man, seated and leaning forward in a posture similar to that of a motorcyclist. His hands appear to be manipulating controls and his left foot is positioned on a pedal. Something looking like breathing apparatus adjoins his nose. And underneath the capsule containing the young man is the plume of smoke from the kind of thrusters which would initiate a rocket launch.

Translation of the inscription revealed that the young man was K'inich Janaab' Pakal the penultimate ruler of Palenque. The rest of the vocabulary of the Temple and its neighbors remained a mystery.

Ruz Lhuillier's find was hailed as the greatest discovery in the history of Mesoamerican archaeology. However, the interpretation of the image remains a matter of intense controversy.

Many have argued for a great range of interpretations. Many have debunked the description as I have presented it and which so gripped Erich Von Daniken when he learned of the carving. The young man was not an astronaut preparing for a flight, but a king preparing to die. The capsule was not a shuttle or craft but the stylized mouths of two serpents, each open at 90 degrees. The billows underneath the capsule were not smoke plumes from rocket thrusters, they were the beards of the serpents into whose mouths Pakal was falling. Or he was playing the part of a Maize God rising up from the jaws of the underworld. But I couldn't see it. I could only see the astronaut. Through the years interpretations and new mythological explanations of the image only proliferated.

Linda Schele was a university professor and a world-leading authority in Mayan epigraphy and iconography. Her study of Pakal's sarcophagus recognized the motifs surrounding Pakal as representing the features of the Milky Way. This suggests the possibility that Pakal's capsule is on a journey not to or from the underworld but a journey through space.

The sarcophagus lid is not the only item of interest at Palenque. A number of beautiful stone carvings in the city portray Mayan leaders, many of whom are sporting elaborate headgear and jewelry depicting what you or I would instantly recognize as a blue-tooth device, worn in the ear, with a microphone extension to the mouth. They are images that leave the onlooker open mouthed and asking, *"What in the world is going on here?"* We seem to be catching glimpses of some out of place artefacts.

If Linda Schele's reading is right and Pakal really is piloting a capsule through the Milky Way, then he is piloting what the ancient Vedic scriptures call vimanas. Expressed in the world's oldest written language these ancient texts date from 1,500BC to 500BC. Their stories make reference to innumerable details concerning the vimanas, how they looked, their airspeed, the kind of noise they made, how many people they could carry, and even specify the elements used in their propulsion systems which included quicksilver and mica. The functions described correlate with what we would understand as space shuttles. They could fly on Earth, they could carry people to a mother ship and make jaunts around the local solar system.

There is no doubt that to the reader delving into ancient Indian history vimanas are presented as an ancient technology belonging to the *"gods"*.

It is curious that we find such technology referenced in literature spanning the globe from Mexico to India. Carvings and artefacts representing this kind of technology can be found among the artefacts of the ancient Olmec, Aztec, Mayan societies, as well as in Egypt and Mesopotamia. Figures sporting costumes that evoke spacesuits, with helmets and breathing apparatus, have been found depicted by ancient Japanese, Chinese, Mayan, Aztec, the Hopi people of North America and by ancient indigenous Australians in the Kimberly region of Australia. The National Museum of Guatemala has a whole section devoted to ancient carved heads with space-helmets. The technologies referenced in their art and costumes are instantly recognizable to the museum's visitors.

But where is any of that in the Bible?

Ezekiel lived from the 6th to 7th century BC and spent a portion of his life in Babylonia. He is known as a writer of apocalyptic prose. *"Apocalyptic"* is a name scholars apply to things seen by ancient writers which are described by analogy. Mind-stretching metaphor is reached for because the writer does not understand what he has seen, but is compelled to describe it, leaving us with images over which the reader also has to puzzle.

In his first chapter Ezekiel describes a mind-boggling encounter. He tells us where it happened – by the River Kebar in Babylonia. Coming down from the sky, Ezekiel witnesses an immense cloud of smoke, filled with light and with lightning sparks emanating from it. He describes metallic legs, burning thrusters, four metallic wheels intersected by perpendicular wheels, enabling the vehicle to manoeuver without having to turn the wheels. The craft is covered in a sparkling metallic or glass-like canopy. The sound of the craft was like the sound of a waterfall. When the engines are switched off the wings lower themselves. From the top of the craft a figure speaks to him. Ezekiel describes him as human-like. He had the appearance *"like that of a man."*

As the craft then carries Ezekiel to Tel Abib, Ezekiel is fascinated by the environment he is now sitting in, he refers to it wonderingly as *"Yahweh's habitation"* and to the vehicle as a whole as the *"glory"*. All through the being's conversation with him in the *"glory"* Ezekiel remains distracted, preoccupied with the swishing sound of the wings, the rumbling sound of the wheels and the loud rumbling noise behind him. When he arrives in Tel Abib, Ezekiel says he found the Exiles living there and *"sat among them for seven days – overwhelmed!"*

Now that is understandable!

Over the next eleven chapters Ezekiel describes similar encounters as the *"glory"* lifts him up into the air and carries him to different places. Each time, while the human-like beings are communicating with him, Ezekiel remains fascinated by how the craft moves, how the wings and the wheels work.

Because of the human-like beings' interest in the people of Israel, we naturally read this as an encounter with the Jewish God – with Yahweh. Ezekiel, too, interprets the encounter in those terms. And it is these experiences which change Ezekiel's viewpoint forever and put him into the role of a prophet to his countrymen, suffering the pressures of exile.

But in all this we are left puzzling, *"Does Yahweh really need a noisy, smoky vehicle to travel in?"* Together with Ezekiel we are left a little overwhelmed, asking ourselves, *"What in the world was that?"*

Somehow our programming bids us to move on and to reason our memory away. Whether we are looking at prehistoric bluetooth devices or ancient space shuttles or beings in space-suits or humanoids in smoky flying machines, we somehow convince ourselves, *"I did not see what I just saw. And neither did you!"*

Anomalous artefacts of planet Earth and anomalous phenomena in outer space offer us no shortage of opportunities to see things which we either have to explain away or unsee. In recent years our viewing of the universe around us has revealed objects which appear to be non-natural. One of the most intriguing is Iapetus – one of the moons of Saturn. Anyone familiar with the canon of Star Wars movies will have no problem recognizing it. The *"Death Star"* is a moon-sized object, spherical in shape, save for two orographic features – a raised rim, marking the equator where the two hemispheres had been welded together, and a large, satellite-shaped crater indenting about a tenth of its surface. The Death Star, in case you're not familiar, is a weaponized mothership – a craft designed to carry millions of people through the reaches of interstellar space. Picture the Death Star and age it a few thousand years and you have Iapetus.

The similarity is so startling it is even acknowledged on the NASA webpage devoted to the moon in question. The massive satellite-shaped crater sits in the same spot just above the moon's equator. Which is curious in itself. But it is the three-mile-high ridge around the equator that is particularly fascinating. It gives the appearance of two planetary halves, welded together. It's hard to see it any other way! On closer inspection the surface of Iapetus turns out to be not spherical. Rather it describes the more geometric form of a dodecahedron.

If George Lucas had created the images for Star Wars in the last decade we might wink at his imagination and creativity in spotting the eccentric form of Iapetus and recreating it as the Death Star. However, what makes this story far more curious is that the Death Star was designed more than a generation before any of us had seen any image of Iapetus. Star Wars was released in 1976. It was 2005 before any of us realized that the Death Star had a real world doppelganger. Perhaps others had seen Iapetus before? Could this be another instance of sharing as fiction what we feel we cannot speak as fact?

Extraterrestrial mother-ships and personal space shuttles have been the stuff of movie lore for much of the movie era. In world literature Hindus have been reading about space-faring vimanas in scriptures which, in written form, go back three to four thousand years – and in oral tradition into pre-history. Similarly the ancient Sumerian narratives tell us of space-stations and the Sky People's mother-ship of Nbiru.

By contrast, Christians do not generally expect to be reading about extraterrestrial technology or space-faring vehicles in the Bible. That was certainly the bias of Josef Blumrich. Blumrich was a senior engineer who served NASA at the Marshall Space Flight Centre as the chief systems designer in NASA's program development office.

In 1972 Blumrich attended a lecture in which NASA had invited Erich Von Daniken to speak. In his lecture Von Daniken addressed the texts of Ezekiel that contain these enigmatic references. During refreshment time, after the event, Blumrich sought the speaker out. *"Mr. Von Dankien,"* he said, *"I enjoyed your presentation but I think you're looking in the wrong place for technology. The Bible is really focused on spiritual matters. It isn't the kind of literature you're going to find anything technical in."*

Von Daniken's simple reply was, *"Have you read the book of Ezekiel?"*

Blumrich promised that he would and applied himself to

the Hebrew text with a view to replying to Von Daniken with a better informed interpretation. However, when Blumrich did sit down with the texts of Ezekiel something else happened. Out of curiosity he applied his engineering skills to the descriptions laid out by the prophet in his book. He began to draw schematics of what was being described. The resultant schema blew all his preconceptions about the Bible out the window.

The result of Blumrich's studies was a book, published in 1974. It was titled, *The Spaceships of Ezekiel.* The subtitle read, *"Was earth once visited from outer space? Did alien beings walk our planet? A major NASA engineer reveals some astonishing facts."*

With a twenty-first century frame of reference, and helped by the work of NASA's Josef Blumrich, today we are in a position to reassure the ancient Hebrew prophet. *"Yes, Ezekiel, you really did see what you thought you saw!"*

According to the Vedic scriptures the vimanas were the technology of the gods. There is plenty of other anomalous technology in our distant past that might leave us puzzling.

In Egypt in 1995, an x-ray was applied to the mummy of an Egyptian leader known as Usermontu. His body was unremarkable, save for the 23cm orthopedic iron pin with which ancient surgeons had repaired his damaged knee. The pin was a flanged screw, held in place by a kind of resin very similar to the bone cement that surgeons use today. Usermontu died around 400BC. Clearly our ancestors were cleverer than we thought.

There is, in fact, no shortage of evidence that *Homo sapiens* have always been *sapiens*. That's to say, we humans have always been clever, and creative and technological. But how did that order of technology get lost? Why did it take two millennia to recover?

Artefacts from Egypt show evidence of stone-cutting that would require power and diamond-grade cutting equipment. Where did that technology come from and where did it go?

Archaeological sites around the planet beg the same question.

In Lebanon's Beqaa Valley lies the ancient city of Baalbek. Around 15BC the Roman occupation began to construct a temple to Jupiter on the foundations of a pre-existing structure. In its heyday the new temple would have been an awe-inspiring demonstration of Roman civil engineering. Yet it is what lies in the pre-Roman layers of the complex that poses the real historical challenge.

In the temple's western retaining wall, four courses of stones up, at a height of around 7m above ground level, is a run of three enormous foundation stones. The stone blocks measure 21m by 4m and 3m thick. They weigh up to 800 tons each. To put that in perspective, a Boeing 747 weighs around 400 tons. The blocks had to be quarried, carried to the building site, and then cut with such precision that you cannot fit a piece of paper in the joins from one block to the next.

Recently, working with a team from the German Archaeological Institute, Jeanine Abdul Massih, of the University of Lebanon, discovered a single stone within the complex weighing one thousand six hundred and fifty tons. The weight of four Boeing 747s!

In Peru the enormous stones of the pre-Inca walls of Sacsayhuaman give the appearance that they have been softened and squeezed into one another to provide a flawless, irregular jigsaw, totally absent of cutting marks and patterned with fluid and curved joins.

Where did the technology come from that enabled ancient engineers to manipulate blocks of stone like that? And where did that technology go? We see the fruits of their technology but where is it now? Who was building these structures thousands of years in our past?

Some researchers, such as the Swedish designer Henry Kjelsen, have speculated that these kinds of feats of civil engineering may have been achieved by means of sound energy. It is a technology we are just learning – or could it be re-learning – in the twenty-first century. In the last decade experiments conducted at research facilities, including Harvard University and the Swiss Federal Institute of Technology, have demonstrated how sonic standing waves can be created which can levitate small objects. There are also reports of this technique being used by Tibetan monks in the early twentieth century for moving much larger objects. From a purely physical point of view, if standing waves can be created today, there is no reason why they may not have been created in the past.

Of course, the sound energy hypothesis is a difficult claim to test. One hint that advanced sound technology may have been part of our prehistory is to be found in a grid of ancient stone circles adorning the landscape of southern Africa.

In recent years, South African researcher Michael Tellinger has made a study of the sonic qualities of the stone used in the stone circles and, in collaboration with other researchers, has measured the sound properties of some of these mysterious stone structures. Even in their current state, which looks like the ruin of an international stone maze, each stone circle is generating phenomenal sound energy.

Tellinger reports that outside the circles is just ambient noise. Inside the circles sound frequencies of 14GHz – that's higher than a dog can hear – have been measured to a magnitude of 72dB – that's between the volume of a vacuum cleaner and a freight train! Curiously the subterranean ground temperature underneath the circles marks the presence of these energy vortices in a measurable way. Outside the circles the subterranean temperature his colleagues have measured is around 5.5 degrees. Inside the circles it ranges from 29 to 58 degrees.

At present these energy signatures are only a matter of intrigue to us. They are anomalies that we can measure. But what can we do with them? Moreover, what was the original purpose of this huge sonic energy grid? And it is immense. The work of researcher Jan Heiner has revealed that this grid extends over a vast geographical area transgressing the modern borders of South Africa, Zimbabwe and Botswana. Could this prehistoric energy matrix have been used as a transportation grid in the ways we are now exploring? And if so what was being transported – and when?

To get an idea of the *"when"* Michael Tellinger has studied the erosion patterns on the stones where they have broken in their current setting. These indicate a timeline for the stone circles that goes back far beyond the conventional timelines of human development. Tellinger's findings demonstrate that the patinas of the rocks give an age of 200,000 to 300,000 years. Who was here hundreds of thousands of years ago to engineer a power grid based on sound energy?

In 2018 the magazine *New Scientist* published a timeline which mapped landmark moments in human pre-history. The earliest built shelters in Japan have been dated to around 500,000 years ago. The earliest evidence of hunting with spears appears around 400,000 years ago. The oldest surviving human footprints (which are universally acknowledged as such) are found in Italy, dating at around 325,000 years ago. The first complex blades and grinding stones appear 280,000 years ago. 230,000 years ago Neanderthals suddenly appear. Then in Ethiopia around 195,000 years ago the first Homo Sapiens have been found. Who was here in the Stone Age, 200,000 years ago, manipulating sound energy? And where did that technology come from?

Clearly there is a piece of the puzzle missing from the story we have told ourselves concerning human origins. Our planet's heritage of built structures argues with paleontological findings concerning the evolution of our species. Geological anomalies, forgotten tunnel and cave systems, submerged and abandoned cities; all point to cataclysmic punctuations in the history of civilization and inexplicable recoveries. Anomalous technologies hint at assistance either from the hidden remnant of older civilizations or from somewhere further afield.

So who has been helping us?

If we allow ourselves to be too wedded to the neat and tidy story of human evolution like the one I read at school, then every time we look upon an anomaly we will be tempted to walk away saying, *"I did not see what I just saw. And neither did you!"* However, if we take time to pause and ponder, the questions begin to pile up and point us in new directions.

My discoveries in the Sumerian tablets made me think again about some of the things I had seen: repeating patterns from one ancient culture to the next, the shared cosmological knowledge, the same evidence of now lost technologies, and timelines that argue with almost everything we think we know about human origins. So I was not totally surprised by what Al Worden had to say on breakfast time TV one Friday morning.

Colonel Alfred Worden is a tough, no BS, NASA astronaut who piloted Apollo 15 on its mission to the moon in 1971. While his colleagues were walking on the moon's surface and trying out the all new moon rover, Al famously orbited the moon seventy-five times.

His remarks came towards the end of his appearance on the UK's *Good Morning Britain* in 2017.

In the final two minutes of the interview, the host, Ben Shephard casually asked him, *"Do you think there are extraterrestrials out there, Al?"*

This was the Colonel's reply:

"You know I've been asked that question hundreds of times; 'Do you believe in aliens?' And I say, Yeah! They say, 'Have you ever seen one?' I say, Yeah…I look in the mirror every morning!

"Because that's what we are. We are the aliens. We just think they're somebody else. We are the ones who came from somewhere else. Because somebody else had to survive and they got in a little spaceship and they came here. That's what I believe. And if you don't believe me, go get books on the ancient Sumerians and see what they had to say about it. They'll tell you right up front."

CHAPTER TEN

LOVING THE ALIEN

This is not how astronauts talked when I was a boy!

When I was growing up any talk of extraterrestrial species or ET contact was confined to easily debunkable encounters reported by witnesses who could be easily dismissed, ridiculed or ignored. At least that's what I saw on the TV. I know now that there was a reason for this.

From the 1940s until around 2008 it was common for governments around the world to employ officers whose remit was to martial reports of unidentified craft and aerial anomalies. The job of these UFO officers was essentially to filter the reports. Those encounters and phenomena which could be easily explained would be released. The other cases – those where an ET explanation was very difficult to avoid – would be classified.

This filtering arrangement is now public knowledge because, from the turn of the new millennium, countries around the world began vacating those departments. Then, beginning in 2008, those who had previously staffed the filtering departments were allowed to form an international body called, The Disclosure Project, which in the years since has campaigned publicly for declassification of all government and military UFO files. If that were not surprising enough, over the next five years a huge volume of previously classified government material was released – case files detailing phenomena, publicly witnessed, filmed, examined by military and civil authorities; cases where an ET explanation was simply unavoidable. All these case studies are now in the public domain.

By contrast the USA has not yet participated in this exercise of declassification. Instead, the USA still enforces the provisions of its seventy-year-old National Security Act, signed by President

Truman in 1947. It provided for the creation of the CIA and simultaneously classified all UFO investigation in the wake of UFO incidents that year at Maury Island, Washington and Roswell, New Mexico.

Yet even in the USA there has been a palpable change in climate. Old patterns of official debunking and ridicule appear to have receded, which is why Colonel Al Worden is only one of many NASA personnel who now feel free to speak openly about ET encounters and phenomena. The number of NASA personnel who have now attested publicly to ET contact include (to name just a few) Mercury Astronauts, Gordon Cooper, Scott Carpenter, and Donald Slayton, Apollo astronauts Eugene Cernan, Colonel Al Worden, Colonel Buzz Aldrin, Dr. Brian O Leary, Edgar Alan Mitchell, and Lt. Col Onizuka. That's quite a body of credible witnesses.

The darker side of what has emerged into the open is an awareness in retrospect of the cruel psychological pressures previously placed upon witnesses and the families of witnesses – a regime of violent gag orders and death threats – particularly following the famous crash incident in Roswell, New Mexico in 1947.

Dr. Edgar Alan Mitchell, the sixth man to walk on the moon, grew up in Roswell. A man of transparent honor, intelligence and good character, he found himself sought out by numbers of *"the old timers,"* as he called them, who were anxious to tell him their families' stories and share their experiences. Dr. Mitchell spoke innumerable times on camera concerning his desire to see government disclose what he believed to be at least seventy years of ongoing contact. His motivations were firstly to see what new chapter in human economics and politics would be possible through the availability of zero point energy and free energy which he believed to be at our disposal as a result of contact. He also wished deeply to honor the families of his fellow townspeople, who in previous years had been silenced through the era of gag orders and threats against themselves and their families.

This story arc tells us of another, more sinister kind of forgetting. The kind of forgetting which is forced upon people when authorities lean over them and say, *"We did not see what you saw. And neither did you."*

I applaud the new climate in which NASA personnel can speak openly and without fear. I think the shift can be read in a couple of ways. It could be that at the beginning of the twenty-first century governments around the world simply agreed together that no one really cares anymore. Alternatively it could be that the change reflects a passive acceding to history, a policy of soft-disclosure. From where I was looking, the new soft policy began to show up in some surprising places.

For instance, in 2011 there was a very curious public spat between the USA and the British Government. The issue revolved around America's demands for the extradition of a computer-hacker by the name of Gary McKinnon. In his leisure, McKinnon had managed to view and extract data, images and text from NASA computers which appeared to indicate a level of ET contact and collaboration.

The matter was discussed publicly in parliament – you can even watch it on YouTube. And the British Government responded with an uncharacteristically hard line towards their international ally. Britain refused to extradite Gary McKinnon to the USA, where he was being threatened with a potential sixty years in jail. In fact, the British Government went further by passing a bill through parliament to alter the UK's extradition laws with the USA in order to protect Gary McKinnon.

It was a bit surreal watching the very public discussion of this conflict play out in parliament. The reporting of NASA's images of ET craft and names of ET officers – without any official comment ever being made as to the credibility or implications of the data itself – was even more curious.

I might have paid little attention to these winds of change in the public discussion of ET hypotheses if it weren't for a move from the Vatican which I definitely did not see coming. In May 2008 a Roman Catholic Jesuit Priest by the name of Fr. Jose Gabriel Funes began writing and speaking to the media about the possibility of ET contact. This would have been unremarkable if Fr. Funes had not been a Senior Vatican Theologian and Director of the Vatican Observatory. The substance of Fr. Funes' statements was that Christian believers need to be ready "sooner than anyone anticipates" to love our extraterrestrial brothers and sisters.

To me, the language Fr. Funes used of making theological room for "a brother alien" indicated that the Vatican was preparing us not for a find of bacteria under a rock or a patch of algae on a distant moon. They were preparing the faithful for contact (or disclosure of contact) with other civilizations.

If that weren't enough of a theological bombshell, Fr. Funes' press release turned out to be only the prelude to something even more public. In 2009, during his brief tenure as Pope, Benedict XVI called upon the Pontifical Academy of Sciences to convene an international colloquium specifically to discuss the theological implications of contact with extraterrestrial civilizations. I had never heard the like!

The colloquium of thirty scientists and theologians met for a five day closed-room session resulting in a formal press release on behalf of the Vatican in November 2009. In the words of Fr. Funes: *"[Astrobiology] does not conflict with our faith because we cannot put limits on the creative freedom of God. To say it with St. Francis [of Assisi], if we can consider…earthly creatures as brothers or sisters why could we not speak of a 'brother alien'? He would also belong to the creation. As a multiplicity of creatures exists on earth so there could be other beings, also intelligent, created by God."*

The Vatican's announcement made me wonder if Rome might be expecting a "sooner than anyone anticipates" disclosure from other authoritative sources and wanted to get in ahead of the game to reassure the faithful. For a colloquium to be convened with a remit like that, and for statements to be released that were as bold as that – all under a Pope as conservative as Benedict XVI – signaled a dramatic departure for the Vatican. After all, it was only a short four hundred years ago that the same authorities were burning people at the stake for merely suggesting that intelligent species might inhabit other planets.

If I had read the cuneiform stories before these climatic changes I would have been less inclined to give them the same hearing. This is because the credence we give to data has something to do with the merits of the data in question and a whole lot to do with our control beliefs.

If I believe that we are alone in the universe, then as a believer I might accept the reality of entities affirmed by my faith community – God, humans, angels, demons – but I automatically rule out any other species or intelligence. If we believe there is only us then we have to view as fiction any theses of plural *elohim,* ET *ben elohim,* Sumerian Sky People or ET interference in human progress. We have to rule out the notion of memory being held in our ancient mythologies. Our control beliefs rule that out.

Control beliefs tend to shift slowly. It's true for an individual and even truer for a culture. So it was only a moment in such a paradigm shift when, on May 9th 2001, more than sixty military, government, corporate, civil aviation and scientific witnesses gathered at the National Press Club in Washington DC to present on their personal knowledge of ET contact and ET-related technologies.

Many of the witnesses were men in their senior years who were breaching their security agreements and protocols and decades of official silence as they testified in front of a major media presence. A number wept as they spoke for the first time of their knowledge and overcame the power of orders, threats, and programming that had kept them silent for decades.

For our cultural worldview to shift and for control beliefs to unravel takes courageous moments like these and a great many of them. Such moments also need the oxygen of publicity. The 2001 disclosure event was attended by a number of major media outlets who all chose not to report the event. However, other agencies were willing to publicize the event – among them, a member of the Vatican Curia – the governing body of the Roman Catholic Church. His name is Monsignor Corrado Balducci – a close friend of Pope John Paul II who served the Archdiocese of Rome as a senior theologian and exorcist. So the public statements he went on to make following the disclosure event of 2001 were significant. He said, *"[ET encounters] are not demonic. They are not due to psychological impairment. They are not a case of entity attachment...These encounters deserve to be studied carefully."*

Monsignor Balducci's statements were soon followed by statements from another senior figure in Roman Catholicism – Revd. Dr. Guy Consolmagno – a senior astronomer at the Vatican Observatory at Mount Graham, Arizona.

Speaking in Harper's magazine in 2006, he addressed the question of how we interpret our sacred texts and what kinds of entities we might expect to find in them. He quotes Jesus' enigmatic words from the Gospel of John, *"I have others who are not of this fold. I must bring them also."* Dr. Consolmagno goes on, *"There are unquestionably non-human, intelligent beings in the Bible...Any creature of this universe, created and loved by the same God who created and loves us...Would they deserve to be called alien?"*

The conversations of the twenty-first century permit us to ask different questions than in times past. In this new climate of soft disclosure I could see too much agreement to dismiss the Sumerian legends as pure fiction. Now I could permit myself to ask whether the Sumerian and Babylonian mythologies might actually be the vehicles of ancient memory, shining a light on the nature of our plural *elohim.*

If the reverend doctors Funes, Balducci and Consolmagno are right and we and our interstellar neighbors really are family, then how exactly are we related? Should we be thinking in Fr. Funes' language of *"a brother alien"*? Or are we, in the words of Erich Von Daniken, *"children of the extraterrestrials"*? Or is the combined chorus of Colonel Al Worden, Plural Genesis and the Mesopotamian and Mesoamerican mythologies on point in identifying us as the aliens – a hybrid race engineered by ET half-siblings?

Maxim Makulov and Vladimir shCherbak believe they know the answer. The two scientists from the al-Farabi Kazakh National University and the Fesenkov Astrophysical Institute have devoted thirteen years to the Human Genome Project, mapping human DNA coding. Together they have concluded that human beings were intelligently designed with what they describe as *"arithmetic patterns"* and *"symbolic language"* encoded into our DNA.

The team published its findings for peer review in the international science journal *Icarus*. Their conclusion is that 97 percent of our DNA's non-coding sequences – what in the past has been colloquially described as *"junk coding"* – is in reality not junk at all but is in fact genetic code from non-terrestrial life forms.

If that sounds like *"alternative science"* Makulov and shCerbak are far from alone in their position. Neither are they on the fringe of their field. In fact the two are among the highest authorities in contemporary DNA research. The bare bones of their thesis had been argued by another significant figure in DNA research as far back as 1981. The paper that presented the case was co-authored by chemist Leslie Orgel and physicist-biologist Francis Crick.

Francis Crick is a familiar name, being one half of Watson-and-Crick, the scientific partnership awarded the Nobel Prize in 1962 for their discovery of the double helix structure of DNA. So these are not the opinions of flakes. They are authoritative voices. And by way of an interesting footnote, the journal in which Orgel and Crick published their pioneering paper on ET seeding was edited at that time by our friend Carl Sagan!

In Carl Sagan's novel, *Contact* the sign proving the intelligent alien authorship of the interstellar signal was the repetition in the code of prime numbers. This was a wake-up signal. The ET intelligence knew that prime numbers would get our attention because they are not numbers that ordinarily occur in natural patterns.

Makulov and shCerbak's evidence is that the prime number signature has indeed been broadcast to us from another intelligence – but not from outer space. The code littered with a prime number signature is our genetic code.

Throughout the genetic code of human DNA numbers keep cropping up that are multiples of 37. The Russian physicist Yuri Rumer first identified one set of repetitions back in 1966. Makulov and shCerbak have identified 9 multiples of 37 throughout our code. Speaking to *New Scientist,* Makulov described the pattern as *"very hard to ascribe to natural processes."*

If that repetition of prime multiples doesn't strike you as odd, Makulov and shCerbak point out that the chances of such a repetition is 1:10 Trillion! With a mastery of understatement Makulov explained, *"It was clear right away that the code has a non-random structure."*

Accordingly, the pair have concluded from their thirteen years of research that the sudden boom in evolution experienced on Earth billions of years ago was not a matter of chance mutation. Makulov writes, *"Sooner or later…we have to accept the fact that all life on Earth carries the genetic code of our extraterrestrial cousins and that evolution is not what we think it is."*

This DNA evidence brings credibility and a finesse to the fifty-year-old story of panspermia, in which some other intelligence seeds the Milky Way with the genetic codes for intelligent, biological life. Popularized in the sixties by Erich Von Daniken, the theory clearly inspired Carl Sagan's *Contact,* as well as the movie lore of Ridley Scott's *Prometheus* and Nick Meyer's *Star Trek II – The Wrath of Khan.*

In their paper in *Icarus* Makulov and shCerbak have this to say about the interstellar transmission of the coding for intelligence:

"Once fixed, the code might stay unchanged over cosmological timescales; in fact, it is the most durable construct known... Once the genome is appropriately rewritten the new code with a signature will stay frozen in the cell and its progeny, which might then be delivered through space and time...It represents an exceptionally reliable storage for an intelligent signature."

If Crick, Orgel, Makulov and shCerbak are correct in their theory, then our intersection with at least some of our interstellar neighbors may be less a case of chance discovery and more a case of seeding and follow-up!

The theory of panspermia carries sufficient weight among the scientific community that some serious money has been invested into its exploration. As a test for the theory, the European Space Agency launched a probe in March 2004 to study a comet with the catchy name 67P / Churyumov-Garesimenko. The probe's assignment was to make the first controlled landing on a comet and look for signs of DNA. The name of the probe was Rosetta – in honor of the Rosetta stone, the historic codebreaker for the hieroglyphs of ancient Egypt.

On November 12th 2014, after a decade of patient pursuit, Rosetta landed on the comet's surface and began its prospecting work. Using a mass spectrometer the probe detected the presence of phosphorous and the amino acid glycine. Both are crucial to the structure of DNA, protein and cell membranes. A jubilant scientist on the project by the name of Matt Taylor told the press, *"This means that comets contain an awesome cocktail of organic material that, if provided with the right conditions, could then go on to form life."*

Panspermia suddenly had wheels!

Movie lore tells another, more cinematic, version of the story. If you have seen Ridley Scott's *Alien* or *Prometheus,* or James Cameron's *Avatar* or *Passengers,* you will be familiar with the idea of sending people or creatures on immense journeys through space in an interstellar ark. Walking around the ship we find the people cryogenically frozen or deep in a state of hyper-sleep, each in their own stasis pod to be opened on arrival. We may even find other creatures or clones growing in their artificial gestation pods, ready to hatch when the ark finally reaches the planet for colonization.

You may think that twenty-first century script writers have really pushed the boat out with these kinds of imaginings. It may sound like a re-writing of Noah's Ark on steroids. In fact these interstellar stories re-sound the notes of another ancient mythology. It is one of the most original and oldest of creation myths, the Zulu legend of Unkulunkulu.

The tale of Unkulunkulu paints a beautiful and evocative scene – a powerfully cinematic version of panspermia. According to the Zulu legend the human story begins…

…When there was nothing but darkness and the Earth was a lifeless rock. From out of the darkness the being known as Umvelinqangi sends a seed to Earth. Within the seed is the life from which all life on Earth has descended.

The seed lands in the soil and sprouts into a bed of reeds full of seed pods. The bed of reeds is called Uthlanga. In one of the seed pods there grew a man called Unkulunkulu – the first ancestor. At first he is a tiny speck. Little by little he grows and forms until Unkulunkulu is so large and heavy that the pod falls off the reed and bursts open.

As Unkulunkulu walks around he finds other people in their seed pods. Unkulukulu opens up their pods and releases them. They are the first humans and the ancestors of all the nations of the world.

As Unkulunku continues his walk around Uthlanga he

finds many forms of life growing in their own seed pods. Breaking their pods open Unkulunkulu gathers the fish and throws them into the rivers. He gathers the birds and antelope and releases them into the wild. He corals the cattle and sends the predators into the jungles and plains.

But the humans are not entirely alone. To help the humans progress from subsistence living to prosperity, a female entity by the name of Mbab Mwana Waresa arrives and teaches the humans how to farm and how to make beer.

I love that detail. Beer is the product of cultivated grains. So beer and farming go together. The Zulu praise Mbab Mwana Waresa for these great gifts to humankind.

However there is a dark element to this story too. Like the story of the Efik, the Popol Vuh, the cuneiforms, and Genesis, the Zulu story tells of death and of making humans mortal.

Unkulunkulu sends a Chameleon to tell the humans *"Humans will not die."* But the Chameleon was too slow. It is beaten by a lizard who is smarter and faster. Unfortunately the faster reptile comes with a different message, *"Humans will die."* From that day to this humans have been mortal.

In this last part of the story we hear an echo of the conflict between Enlil and Enki over the plan to genocide humanity by means of the flood. It is echoes of the agreement of the Sky People and the Powerful Ones to limit human lifespans and make humans mortal. Both elements are in the Biblical narrative too.

The Zulu legend holds many layers for the Zulu people and I would not wish to trespass on all those beautiful and sacred layers. But the motif of life on Earth – the seed form of plant, animal and human life – arriving on Earth from another place is a story too powerful and too resonant for the wider world to ignore. And might there be a reason that contemporary story-tellers bring us this telling so many times over? Fiction resonates the longest when it speaks a truth about ourselves. Could the Zulu legend be a case in point?

Of course, ultimately none of this speaks to the question of the original source of life – only the way in which life and intelligence might migrate through the universe.

Makulov and shCerbak's vision is not of a spaceship with stasis pods but of objects such as comets carrying the DNA coding of the parent culture through the galaxy. The senders have disseminated the coding into the Milky Way, knowing that hospitable planets will grow and multiply the life forms encoded in the DNA sequences. In this biological diaspora the senders are literally sowing the seeds of civilizations which, aeons into the future they will be ready to follow up.

My question in the light of Maulov and shCherbak's thesis was exactly what that follow-up might look like? Might it involve a little further tweaking; the kind of tweaking that turns a primate into a hominid, a hominid into a homo erectus and a homo erectus into a human?

CHAPTER ELEVEN

FATHER GOD, I WONDER

"So, Paul, let me get this straight. You're telling me that you are descended from a genetically modified clone of a hybrid of a hominid and an ET? Well by contrast, my old friend, I am very happy to say that I happen to be a child of God – which by my reckoning is a better deal! Can I get you a beer?"

I was at Oldstream Pass once again, chewing the fat with my theological friend, Brad. He went on, *"Somehow I had always guessed that you might be the result of a comet randomly dropping acid into the ocean! But honestly, Paul, you can't believe what you're saying and still call yourself an orthodox Christian. It says in the Creed, 'We believe in one God, the Father, the Almighty, maker of heaven and earth, all that is, seen and unseen.' I think that rules what you're saying out."*

Not really. The millions who hold to the affirmations of the Nicene Creed don't have any problem understanding God's relationship to things that we human beings have created. For instance, our Heavenly Father didn't create the Manhattan skyline, beautiful though it is. God didn't create the city of Hong Kong or the land on which it stands. He didn't build the Sydney Opera House or the London Eye or the Leaning Tower of Pisa.

It's not hard to work out that while human beings engineered these things, none of them could exist without God who is the source of every atom, every proton, electron, photon, energy, matter, dark matter or anti-matter. Indeed when we use the word *"God"* or *"Creator"* we are really affirming the notion of God as the Ultimate Source of all things, the Ground of all being. His is the consciousness and will on which everything else depends and from which everything else emanates.

If we believe God to be the source of life itself then the same logic applies to living things. For instance, my family used to own a beautiful German Shepherd dog called Saba. God did not create the German Shepherd as we know it. Less than a century ago nothing like the modern German Shepherd existed. We have bred them over a century to be a different size and shape of dog to their ancestors. Knowing that we have genetically modified and cross-bred them doesn't lead me to conclude that God didn't love my dog!

"Did you know," I said, *"that new types of bear have been appearing on the planet over the last few years. As Earth's climate warms and our ice-fields gradually melt, they've found that polar bears have been mingling with grizzly bears who have been migrating further north. Result: new kinds of bear! They're calling them "Grolar Bears" or "Pizzlies." If we have influenced shifts in the climate, then that's our work but I don't have any problem believing that our Father loves Grolar Bears and Pizzlies every bit as much as their pure-bred parents."*

The same would apply to Dolly. She was a true pioneer. In fact no sheep like Dolly had ever existed before. All previous sheep – so far as we know – had resulted from a sheep egg fertilized in utero by sperm from a ram. The egg that became Dolly was fertilized in vitro – and not by a ram but by a DNA extract from her mother. Dolly was a clone.

Born in 1996, Dolly, the world's first cloned sheep lived a long and happy life until she died of old age in 2003. Humans didn't create Dolly the sheep *ex nihilo*. We just cloned her. Today we can splice and modify and adapt our cloned sheep to suit our needs – at least as best we know how.

Whatever the Almighty may think of our involvement in cloning, I don't think we have any reason to suppose that God loved Dolly any less because of her parentage? If God is the source of consciousness, life, the will to live, the source of every proton, electron, photon *etc.* then I have no need to see our Heavenly Father as any more remote or less loving towards

Dolly. Similarly if we *Homo Sapiens* really have been genetically modified by our galactic neighbors, or artificially adapted from earlier hominids, does that mean that God, the sustainer of every atom, and the source of life and breath and consciousness, is therefore not our Father?

If God is the source of every atom, proton and electron that carries my consciousness, then I can still address my God and say:

"You created my inmost self, knit me together in my mother's womb…You knew me through and through. My being held no secrets from you, when I was being formed in secret, textured in the depths of the earth. Your eyes could see my embryo. In your book all my days were inscribed. Every one that was fixed is there." (Psalm 139:13–16)

This is something I can affirm whether I was conceived in utero or in vitro; whether my blood is human, pure and simple, or human with a few drops of ET!

Fr. Funes of the Vatican Observatory suggests that believers need to expand their picture of God's parenthood and loving care to cover all of creation – every person – of every species across the universe.

Joining the dots from one creation mythology to the next felt like an invitation to see myself as part of a bigger creation. Peering into the photo albums of our wider family showed me that I was the member of a bigger family than I had ever imagined before.

But where is God in this picture? If the stories of God-as-he-really-is don't begin with the first verse of Genesis, then when does he show up? If we use the revelation of Jesus as our lens then when do we first clearly perceive the God who is the Father of Jesus?

Is it his apparition in the burning bush and his conversation with Moses that introduces the true God into the story?

I had always noticed that Moses was quite confused by this encounter and responds as if he has no idea of who is talking to him. He is baffled by how he is going to explain the identity of this burning bush entity when he returns to his people. This is striking. If Moses is the descendant of Jacob, the descendant of Abraham, how can he not know this God? Clearly something profoundly new is being revealed.

Moses isn't the only one confused by Yahweh's arrival on the scene. If the people of Israel had known that their God Yahweh was the ultimate God, the Almighty, the Source of all things, then their story, moving through the Bible, would have been completely different. Their repeated falls into apostasy and the idolatrous worship of other entities, the Powerful Ones of other people groups, are inexplicable until you realize that early on, the people of Israel did not know how to distinguish their Yahweh from all the other entities they knew about – the Powerful Ones of their neighbors and the Powerful Ones of their Ancestors.

Though Joshua issues the people of Israel a clear call to reject the Powerful Ones and turn exclusively to Yahweh, it's clear that this new monotheistic lifestyle and belief takes a long time to bed down.

Meanwhile their leaders and scribes are chronicling their journeys and their evolving relationship with Yahweh. By the time we get to the minor prophets such as Hosea we are learning that True God loves us – not because of anything we do to make ourselves worthy, but purely because God is love and he loves us. By the time we reach the prophet Amos we are getting to see that True God loves every people group and wants goodness and justice for every human being – even if they should be worshipping the wrong gods! The prophets tell us that True God isn't interested in sacrificial religion. The sacrifice he loves is the expression of love, goodness, kindness and justice towards one another in human society.

The problem is that even after the arrival of Yahweh as a character in the Bible's story the picture does not immediately resolve. It would be easy if there were a clean cut between the elohim stories and the Yahweh stories. But there isn't. Furthermore, it isn't entirely clear if the Powerful Ones of Israel's neighbors are now entities purely of ancient memory, or if they're still around.

Other Powerful Ones appear as present entities as late as the account of Daniel from the Babylonian exile. In the tenth chapter of that book an angel from God visits the Israelite prince Daniel and reports that he had some difficulty getting there in a timely fashion because of having to battle first with another powerful entity, whom he calls the guardian of Persia. Could this guardian be one of the problematic members of the Council? Whoever he is he's evidently a power to be reckoned with.

The question is raised again in the book of Judges when Yahweh tells the people not to be afraid of the Powerful Ones of the Amorites in whose land they are living. If the Powerful Ones were only memories or images, why would he have to say that to a people aided by Yahweh, the Almighty? Why does he address the presence of Powerful Ones as if they were something?

In the book of *I Kings,* King Ahaziah, the King of Judah, falls from an upstairs window and is left bedridden. Elijah, the local prophet of Yahweh, finds out that the king is about to send messengers to consult with the Powerful One in Ekron to find out if he is going to survive. On hearing of this the prophet goes in to the poor king who is lying, broken and in mortal pain on his bed, and, on Yahweh's behalf, blasts the poor king: *"Is there no Powerful One in Israel for you to consult that you should have to send messengers to consult the 'Lord of the Flies', the Powerful One of Ekron? Because you have done this you will never leave that bed you are lying on. You will certainly die."*

It is as if Yahweh and the Powerful One of Ekron are in some kind of competition of the gods. In fact Yahweh even mocks the other god's name. *"Lord of the flies"* is one letter different to his real name which means *"The Lord and Prince."* Furthermore, he uses the same word *"elohim"* to denote both himself and the Lord and Prince of Ekron. This designates them both as Powerful Ones. If the Lord and Prince of Ekron was only a totem or an object and Yahweh is the Almighty, ready and available for Ahaziah, then both Yahweh's and the king's behavior makes little sense. And Yahweh shows Ahaziah no mercy – at least not the way Elijah represents him.

What was King Ahaziah seeing if, even with his knowledge of Yahweh, he took seriously the prognostic skills of the Powerful One of Ekron? Ahaziah's behavior is only one instance of many similar infractions among generations of kings of Israel and Judah. What exactly was the draw of other *"gods"* if other gods were not real and they had the real, living, Yahweh the Almighty with them?

It is a conundrum built into the Ten Commandments of Mosaic Judaism (Exodus 20 and Deuteronomy 5) The first command is:

"You shall have no other gods (Powerful Ones) before me."

If other gods (Powerful Ones) don't exist why does this have to be said? We may rationalize and say, *"Oh he must mean imaginary gods or idols."* Except, that is covered in the very next sentence:

"You shall not make for yourself a carved image—any likeness of anything that is in heaven above, or that is in the earth beneath, or that is in the water under the earth; you shall not bow down to them nor serve them. For I, the LORD your God, am a jealous God, visiting the iniquity of the fathers upon the children to the third and fourth generations of those who hate Me, but showing mercy to thousands, to those who love Me and keep My commandments...." (Deuteronomy 5:8)

The commandment implies a multiplicity of Powerful Ones – all to be rejected with the exception of Yahweh. This mirrors Joshua's call to the tribes of Israel to reject the Powerful Ones of their neighbors and of their ancestors.

There is another layer to this picture. Consulting the Powerful One of Ekron may not have implied a face-to-face encounter between Ahaziah's messengers and a powerful entity. The Ekronites pattern of consulting may have involved forms of divinations, interpreting patterns of sticks or stones, or reading the entrails of sacrificed animals. It may have involved sitting before an idol or totem carved with the images of real and fantastical creatures and allowing a prophet to relay a message from the powerful one represented by the idol.

It may look strange to the modern eye but it is not a million miles from the way in which the kings of Israel and Judah and their prophets consulted with Yahweh. Notwithstanding the instruction to *"make no carved image—any likeness of anything that is in heaven above, or that is in the earth beneath, or that is in the water under the earth..."* Yahweh gives instructions to Moses for the construction of an object to totemize his presence. The *"ark"* or *"tabernacle"*, as it is called, is to be adorned with two modelled winged creatures, with their wings spread upwards over the seat from each side. Whenever a leader wishes to consult with Yahweh he is to sit on the seat between the two images in order to receive instructions from Yahweh.

Seen against the behavior of the neighboring tribes who adhered to other Powerful Ones, it is much more of a par for par image than we generally think. Ahaziah may have been comparing totems rather than entities.

So the presence vs non presence, both of Yahweh and of other Powerful Ones, is not quite as clear cut an issue as we might prefer.

Neither is it a simple matter to identify when and where portrayals of God enter the Hebrew Scriptures which would in any way harmonize with the image of God as seen in Jesus. Yahweh's judgment of King Ahaziah, which I mentioned earlier, pales in comparison to his harsh treatment of King Saul – Israel's first king – a man who did nothing other than try to do right by Yahweh.

In the book of *I Samuel*, King Saul returns from victory in battle. Through the prophet Samuel, Yahweh has sent King Saul to war to punish King Agag and his people, the Amalekites. On his return from battle Saul greets Samuel with the words, *"May you be blessed by Yahweh. I have carried out Yahweh's orders!...The people [have] spared the best of the [Amalekites] sheep and cattle to sacrifice them to Yahweh..."*

But the prophet cuts him off and upbraids him for failing Yahweh. At first Saul is baffled and confused. He says, *"But I did obey Yahweh's voice. I went on the mission which Yahweh gave me. I brought back Agag, king of the Amalekites. I put Amalek under the curse of destruction and have taken the best sheep and cattle of what was under the curse of destruction to sacrifice them to Yahweh your God in Gilgal."*

Samuel retorts, *"Obedience is better than sacrifice!"* and carefully reminds Saul of the small print of his orders to go to war. The instruction from Yahweh had been not just to defeat the Amalekites but to *"kill [every] man, woman, child, nursing baby, ox and sheep, camel and donkey."* Yahweh had wanted a scorched earth genocide. Saul had fallen short. Samuel continues, *"Let me tell you what Yahweh said to me last night...'I regret having made Saul king, since he has broken his allegiance to me and not carried out my orders.'"*

Saul finally understands how he has offended Yahweh. Note the whole conversation is mediated through a prophet, so this accession requires a great deal of humility on the part of the king. When Saul understands his mistake he is distraught, pledges his fealty to Yahweh and effusively begs for his mercy. He pleads with Samuel, saying, *"I have sinned...please forgive my sin and come back with me so that I can abase myself before Yahweh."*

Samuel, however, makes clear that Yahweh is not to be reasoned with and will never forgive him. He rejects Saul as king, anoints a successor while Saul is still on the throne, and sends a demonic spirit to afflict and torment the king. Even aggressed against in this way, Saul still tries to seek out the will of Yahweh through summoning the spirit of Samuel the prophet to bring him Yahweh's word. But Yahweh has cut him off and uses the demonic spirit to drive him into insanity and death through suicide. If that is how Yahweh treats his friends....

Where the king has failed Yahweh the prophet Samuel steps in. *"Bring me Agag,"* he orders. The text continues:

"Agag came towards [Samuel] unsteadily, saying, 'Truly death is bitter.'...Samuel then butchered Agag in the presence of Yahweh at Gilgal."

This is the unsavory story that marks the transition of kingship from divine hands to human hands in the history of Israel. In the Egyptian and Sumerian mythologies their kingship makes a similar transition into human hands. However, in their case, the Sky People hand the reins of power to their human successors quite peaceably. For poor King Saul, Israel's first human king, the transition was a tragedy. It is clear from the book of I Samuel that Yahweh was deeply offended by the people of Israel's decision to replace him with a human king. Yahweh's turning against Saul after the Agag episode really looks like a vindictive act of retribution to sabotage Saul's reign – and all on the grounds that Saul had vanquished the enemy in a war but failed to effect a total genocide. There may be military rationales and precedents for this but it would be a more ingenious preacher than me who could square the morality of this episode with the teachings of Jesus.

This is only one of many stories of Yahweh that challenge our morality. But it's a notable one because in it Yahweh's victim is a man who is doing his best to follow and obey his deity. Elsewhere in the Hebrew Scriptures there is no shortage of reportage of Yahweh's interactions with the tribes of Israel and their neighbors raising considerable difficulties for any reader who measures Yahweh's conduct against the revelation of God in Jesus.

II Peter teaches that to participate in the nature of God requires a person to put on goodness, wisdom, self-control, patience, godliness, brotherly kindness and love. Apostle Paul tells us that love is kind, is not proud, does not envy, does not boast, is not rude, or self-seeking. It is not easily angered and keeps no record of wrongs. There are God-stories in the pages of the Hebrew Canon which present a character that would appear to be at odds with all those descriptors. They raise the question of how quickly and clearly the worldview of the tribes and scribes of Israel moved from polytheistic roots to monotheistic clarity.

In the time of Moses, Yahweh speaks through a fire, and then descends on Mount Sinai amid fire and smoke. Similarly when Yahweh arrives to collect the prophet Elijah, he turns up rather dramatically in a vehicle that descends from the sky, belching out clouds of smoke and jets of fire, picks up Elijah and flies up into heaven through what we might describe as a wormhole. Elijah's vehicular departure is strangely reminiscent of the prophet Ezekiel's other worldly experience.

Does an almighty, transcendent God really need a smoky vehicle to move around in? Does he really need the device of a totem or ark through which to communicate? Where a worldview rooted in experiential interactions with Powerful Ones ends and a clear vision of True God begins in the Bible's revelations is not crystal clear.

In the beginning we read a sequence of stories of the Powerful Ones. Towards the end of the Hebrew Canon we have revelations that are largely congruent with what we see and hear in Jesus. But in the middle we are confronted with no shortage of behavior accredited to Yahweh which is very hard to square with the God seen in Jesus. One could mention the cruel and unusual treatment of King Saul in *I Samuel 15*; or the killing of 70 Israelites for being terrified by the power of the ark, which had caused tumors in their Gathite neighbors in *I Samuel 6*. Or we could consider the mass-murder of 70,000 innocent Israelites under the reign of King David in *II Samuel 24*. All these actions are attributed by the narrators to Yahweh. These attributions are baffling. Certainly they are disturbing when judged by any moral standard. When measured against the qualities of God's nature as extolled in the New Testament they simply don't compute. This is the problem that Origen and Marcion squared up to in the early years of the Christian faith.

Whether Christianity would simply import contemporary Judaism's reading of the elohim stories as God stories was not a foregone conclusion. Even within the pages of the Gospels and the book of Acts we can hear the push and pull of debate as the early Christians decided how much of Judaism Christianity needed to affirm. It took decades, even centuries, for the early Christians to work out what it meant to be an international faith with an international Savior, who just happened to have sprung from Jewish roots. Just as Mosaic Judaism had to reject the religious baggage brought by its patriarch and matriarch, Abraham and Sarah, (ie "the elohim of your ancestors") might International Christianity have to shake off the religious baggage of the Jewish faith from which Jesus had sprung. The answer was "yes". Accordingly, by the fifteenth chapter of Acts a general council of Christianity's leaders, chaired by Jesus' brother James, concluded that the Jewish law no longer held force for Christian believers.

That left the question of what to do with the Hebrew Scriptures, with their morally dissonant depictions of God. Marcion was not alone in proposing that they be left to one side. Justin Martyr and Clement of Alexandria – both significant church leaders in the C2nd – both argued that the Hebrew Scriptures were the philosophical heritage of the Jewish people. God had spoken through that heritage to prepare the Jewish people for Jesus. (Jesus and Paul both say as much in the Gospels and New Testament.) In the same way, God had prepared the people of the wider world for Jesus through the knowledge it had through Greek philosophy.

The logic of their view is that while Jewish Christians might honor their Scriptures as the means by which they had come to faith in Jesus; that the Gentile / "Greek" world might equally honor Gentile / Greek thought as the foundation of its understanding and belief in Jesus. Along with many of the Church Fathers of the day, Clement understood *"Greek thought"* as being best crystallized in the works of Plato and the Stoics. Mathematically, from the point of view of benefit to the Christian mission, that had to make Greek thought – *ie* Platonism – the more useful platform.

If that logic had been pursued, the Church would never have glued the Hebrew Scriptures to the New Testament writings to create an international Christian Bible. The flavor of Christianity would have been palpably different if Jesus' teachings had been seen in the context of Greek thought rather than Jewish history.

By gluing the Hebrew Scriptures onto the New Testament, Christianity brought within itself the version of human origins that *"J"* created when he redacted the elohim stories and turned them into God stories. It brought within itself a cosmology in which there is only the human race with no E.T. neighbors to complicate the picture. It brought within itself a depiction of God that is confused, morally dissonant and ultimately incompatible with the revelation of God in Jesus. Origen's response is a wonderful illustration of the somersaults Christians have turned to in order to try and avoid the problem ever since.

Put simply, the early church had a choice. They could uphold the New Testament of Jesus fundamentally as its own canon and allow people to read it in the light of mainstream international thought – expressed in Platonism. Or they could define the Christian New Testament more parochially and tie it to the Judaism of the Hebrew canon.

If they had followed Justin and Clement's logic and accepted the challenge of Marcion then Christianity would have taken a different shape, one less at odds with the wider world, one more open to other explanations of beginnings, other mythologies and other species. Jesus' life before incarnating on the planet – his pre-existence as divine thought (logos) – would have been seen as a model for our own origin and self-understanding. And God, the Spirit, the Source of all things, would have been viewed without the need to (secretly) suppress fear and loathing at his violence.

We would have seen a less violent Christianity, one less useful to imperial powers and feudalizing social orders. Indeed, the whole of Common Era history would have played out quite differently.

So I have to ask: by anchoring Christianity to *"J"*'s recasting of the *elohim* stories, did the Church get it wrong, thereby distorting Jesus' revelation of God for the next two thousand years?

The cuneiform tablets and the other creation myths had sent me back to the Bible to wrestle the same wrestle. My questions around the translation of *"elohim"* had shone light on a whole number of God-stories that may not be God-stories after all. With my eyes open to that possibility, I could see why both Origen and Marcion might deserve a more careful re-listen. They were the two theologians who, in the earliest days of the Church, spoke most boldly into these questions.

I put this to Brad.

"Well there you go!" he said. *"You want me to give credence to your crazy views on the strength of a couple of shady figures who, if I remember correctly, both got thrown out of the church as heretics? Face it Paul you're a heretic. Very thoughtful. Carefully considered. Deeply interesting. But a heretic! Don't tell me you'll be preaching ancient aliens at Church on the Range? Because I don't think they're going to like it! Do you want another beer? It's your round!"*

Brad had a point there. As it happened, though, it wasn't going to be an issue. The impact of my Frisbee injury had proven longer lasting than anticipated and had thrown up some unforeseen consequences. My extended time of being twisted and pulled in traction, and the theological torsions of my mythological travels had made me altogether less useful to my flock than they or I might have preferred. So I felt it better to release them from their obligation to me. Hence I was going to have to be a heretic on my own dollar! Which reminded me; I was going to need to vacate my cozy shipping crate sometime soon. Without my stipend from the Church on the Range we were going to be needing that AirBNB money!

CHAPTER TWELVE

I HAVE CALLED YOU FRIENDS

If I had been preaching any time soon, perhaps I might have been tempted to lean towards Origen's tactic and disregard the plain meaning of the text whenever the content got a bit hairy. His advice to the preacher confronted with a *"problem text"* – *i.e.* one in which the narrator asks us to believe monstrous things of God – was to preach on the moral of the story or find in it an allegorical or esoteric reading, prefiguring Jesus or the Church. All very well if it's just a sermon you're looking for but it avoids actually exegeting the text and wrestling with the claims made by the text.

By contrast Marcion was open in promoting the same conclusion I had reached through my mythological travels – namely that not all the God-stories in the Bible really are God-stories.

Marcion had grown up in the Christian faith. He was a bishop and the son of a bishop. From his studies he concluded that the *elohim* of the creation narratives represent powerful entities or a powerful entity that is not God. His reasoning continued that if God is as he appears in Jesus then he cannot possibly be as he appears in many of the Yahweh stories. The two characters are totally different to each other. Marcion's response was to make the revelation of Jesus the arbiter of any truth claim about God – even the truth claims of the Hebrew Scriptures. He therefore excised from his Bible every book that makes contrary claims about the nature, character and behavior of God. For him that meant ditching the Hebrew Scriptures. Among the gospels and letters that were soon to comprise the New Testament, Marcion was a great champion for the Apostle Paul's letters and their associated gospel, the Gospel of Luke. Marcion felt that the

Apostle had genuinely grappled with the Hebrew tradition and transfigured it in a way that was accurate to the Gospel of Jesus.

A significant number of churches throughout the Mediterranean agreed and followed Marcion's approach. However, the mainstream didn't accept it. The problem for them was that Jesus had affirmed the Hebrew Scriptures when he said, *"These are the Scriptures that testify to me"* and *"The Scriptures cannot be broken."*

Accordingly, Jesus and his Apostles went to the Hebrew Scriptures – never for the plain meaning, but for allegorical or esoteric readings. They transfigured the old mythologies into *"types and shadows"* to reflect the new teaching of Jesus.

Jesus did not rubbish the Hebrew Scriptures. Nevertheless, he made very clear what level of authority he felt they held relative to his own assertions of truth. *"You have heard it said...but I say..."* or *"Moses said this but I say this..."* It's clear that the early believers wrestled with exactly where these kinds of pronouncements placed the Hebrew tradition in the new religion. But by *Acts 15,* when the primitive Church called a General Council to resolve the question, it had become clear that Jesus had *"abolished the law with its commandments and ordinances," (Ephesians 2:15)* The imperative of the Hebrew commandments had passed. The Church had moved on.

Nevertheless, a century later Orthodox church leaders joined in a chorus of disapproval against Marcion's impious treatment of holy scriptures – especially the Hebrew Canon. Sadly, the Church hasn't retained any record of Marcion's writings. Somehow they all seem to have got lost or destroyed. Quite a feat, given the wide geographical spread of the Marcionite churches. Today we only have the refutation of Marcion's work by the leading lights of the mainstream. So one has to sift a bit to separate what he actually said from the Aunt Sally version of his position, pilloried by his fellow bishops.

And they didn't pull their punches in theological debate back in those days. Polycarp, the spiritual grandson of the Apostle John, famously called Marcion to his face, *"The first-born of Satan."*

The church father Tertullian made fun of Marcion's approach in his paper *Against Marcion,* saying, *"Listen you sinners...a better god has been discovered, who never takes offence, is never angry, never inflicts punishment, who has prepared no fire in hell, no gnashing of teeth in outer darkness! He is purely and simply good...They say it is only an evil being who will be feared. A good one will be loved...And the Marcionites are so satisfied with these pretenses that they have no fear of their god at all."*

But all this failed to address Marcion's central question of how we deal with the incompatibility of the character of Jesus over and against depictions of Yahweh and the *elohim* in the Hebrew scriptures. The problem for Marcion was not simply that the claims about God were offensive in some general kind of way. It was that if God is as he appears in Jesus then how can he possibly be as he appears in some of those Old Testament texts?

It wasn't until Origen arrived on the scene a generation later that the mainstream churches found a leader willing to confront the same question and provide a way forward for an orthodox mainstream wishing to keep the Hebrew Canon as the Christian Old Testament.

The Church is hugely indebted to Origen. His allegorical approach and his argument for an esoteric reading of problem texts laid the foundation for Christian preaching from that day to this. Indeed, the Church was so grateful to Origen that it even did him the favor of posthumously editing his best works to make them even more orthodox for posterity!

However, what is often understated is that Origen's reasoning begins with the exact same statement of the problem that set Marcion's theology in motion. One sentence from Origen summarizes his analysis of the problem and the bare bones of his response:

"It was after the advent of Jesus that the inspiration of the prophetic words and the spiritual nature of Moses' law came to light." (My underlinings.)

In other words Jesus' appeal to the Hebrew Scriptures affirms them as *inspired*, uses them as a *prophetic* reference and draws *spiritual* meanings from them. At the same time Jesus' revelations make it impossible for us to read those Scriptures at face value. So Origen goes on to say that people make mistakes when *"... they fail to understand Scripture in its spiritual sense, but interpret it according to the bare letter."* As a result readers *"believe such things about [the Creator] as would not be believed of the most savage and unjust of men."*

Origen is very clear that where God is portrayed as *"savage and unjust"* the wise reader must utterly reject the plain meaning – the *"bare letter"* – of the text and find an esoteric or allegorical interpretation instead. My question would be, *"Is that really honest? Does that expound the text or actually cordon it off?"*

At one level Origen's answer is as clear as Marcion's in rejecting the portrayals of *elohim* and Yahweh as pure and unalloyed expressions of the God and Father of Jesus Christ. But it sowed a kind of double-talk into Christian thinking by saying, *"We believe the text to be inspired but deny its claims."* It was a way of assimilating the Hebrew Scriptures – but not really.

As early as the second and third centuries, apologists like Justin Martyr and Church Fathers like Clement of Alexandria were arguing that the world at large had been prepared for the teachings of Jesus by the forms and worldview frameworks of Gentile (i.e. non-Jewish) philosophy or *"Greek thought"* – which they considered to be best expressed in the thought world of Stoicism and Platonism. If the Church were to be a vehicle of Jesus' teachings for the world then, by that logic, one would arrive at a canon of wisdom comprising an Old Testament of Plato and a New Testament of Jesus and the Apostles. However, the hints of Justin Martyr and Clement of Alexandria failed to win the day. The New Testament was, in the end, simply glued to the Hebrew Scriptures to create the Christian Bible.

So the moral dissonance of Yahweh, woven into the current redaction of the Hebrew texts then became dissonance within Christianity, with the result that Christians now had to defend actions attributed to The Almighty *"such as would not be believed of the most savage and unjust of men."* Evidently the Church's vision was of a God to be feared and loved unquestioningly.

But was that the right call? What would Christianity look like if we were to follow Justin Martyr and Clement of Alexandria's nudges towards a Christianity framed by first century Greek thought? What would Christianity look like if we accepted the theses of Marcion and Origen which, each in their own way, rejected the OT's theological claims?

Would a Christianity absent of the Old Testament and framed by the start-point of first century Greek thought produce a clearer vision of God, along with a clearer and radically different vision of human nature and human origins? Would a framing of Christianity that was more friendly to the thought world of the international community have altered the progress of the Church as a demographic within the world?

The context of J's redaction of the Jewish mythologies was a religious imperative to say to their foreign dominators, *"Your gods are no gods at all. Our mythology is the truth. Yours is a lie. Our God is the true God. He will justify us and condemn you. He will vindicate his people and destroy their enemies. Everlasting reward will come to us and everlasting shame will fall upon you."*

By adopting J's stories as the framing for Christianity the Church set itself at loggerheads with the world and framed the teachings of Jesus with a narrative of us and them, right and wrong, love and hate, reward and retribution, heaven and hell, along with an expectation of perpetual persecution.

I have to wonder, would a Christian Bible absent of the Hebrew canon have created a more affirmative relationship with humanity, and a healthier relationship with the world at large?

These were the kinds of considerations that Justin Martyr, Clement of Alexandria, Origen and Marcion all offered the Church in that early period and you can see the answer to those questions in some of their writings. But it was Marcion who ran the furthest with that proposition and who catalyzed the reaction that ruled Greek thought out and the Hebrew canon in.

Could it be that Marcion's proposal of a root and branch break with the Old Testament mythologies was more genuinely in proportion to the problem? There was no pretense about it. Marcion's response showed an intellectual honesty and a transparency that Origen and the voices of orthodoxy did not. I believe Marcion was also right in his assessment that the Apostle Paul had transfigured the heritage of Hebrew Scripture into a typology for the radically different message of Jesus concerning the character of God.

Apostle Paul's influence is without parallel in his establishing of Christian churches and his authorship of much of the New Testament. As a convert to Jesus after a formation in Pharisaic Judaism, Paul often found himself preaching to Jewish audiences to persuade his hearers of Jesus' identity as the promised messiah. In that context Paul often drew upon his Jewish heritage. Yet his visits to the Hebrew Scriptures are essentially esoteric ones in which he somehow unveils previously hidden layers of metaphor and meaning.

So, for Paul, Sarah and Hagar, respectively the wife and concubine of Abraham, represent two covenants. In another instance a law from the book of Deuteronomy about the care of cattle is really an allegorical message about paying church leaders. Elsewhere Paul appeals to the moralization of Adam as a prophetic shadow version of Jesus bringing humanity the gift of eternal life. But on the great bulk of the Hebrew Scriptures, Paul is silent. In that sense we can see the early seeds of both Origen and Marcion in the approach of Apostle Paul.

It is not that Paul's theology is absent of shade but there is no defense of or even reference to the kind of deity who is unpredictable, inscrutable, reactive and punitive, who can turn on a dime and exact plagues and punishments on people doing their sincere best to obey him. Neither does Paul's vision of God see a divinity who needs to be appeased with tribute and sacrifice. Rather Paul presents the person of Jesus as the fulfilment and final conclusion of the Hebrew sacrificial tradition.

In the book of Acts we get to eavesdrop on a speech the Apostle Paul made to a Greek audience in Athens in around 51AD:

"The God who made the world and everything in it is himself Lord of heaven and earth, and does not live in temples made by human hands. Nor is he served by human hands as if he needed anything! On the contrary he is the one who gives to everyone else. Everything is given by him – including life and breath. From one blood – he made all the people on earth, arranging the times set for them and the places where they would live. And he did this so that people might seek the true deity, reach out and feel their way towards him and find him; thought is not far from any of us, since it is in him that we live and move and have our being...' (Acts 17:24–27)

In this speech, for all his fidelity to his Hebrew formation, Paul dispenses with the whole idea of sacrificial religion, ridiculing the idea that the Divine Source of all things would need us to keep him supplied with anything! In speaking this way Paul is not rejecting the Jewish tradition. In fact he echoes the contribution of a number of Jewish prophets through the ages when they spoke for God.

"I don't want your sacrifices. I want your love. I don't want your offerings. I want you to know me." (Hosea 6:6)

"No I don't need your sacrifices of flesh and blood. What I want from you is true thanks: I want your promises fulfilled. I want you to put your faith in me." (Psalm 50:13–15)

In Athens Paul issues the call to a higher consciousness of

the True God in an environment over brimming with shrines and temples, created so that sacrifices could be brought to all kinds of other entities, demi-gods, hybrids and others. In fact Athens was home to the most famous demi-gods and hybrids of the world's mythologies. All were deferred to in the panoply surrounding the Apostle Paul on that visit.

The urge to appease or make sacrifices to gods as a way of prospering in this life and the next is deep wiring that transcends our various cultures. The Mesopotamian mythologies give a compelling explanation as to why this programming runs so deep. They assert that the religious impulse to bring sacrifices and offerings to our higher beings is a behavior rooted in our original wiring as slaves, trained to bring mineral and food supplies as tribute to temples and tithe barns for use by the Sky People and other superiors. It is a cultural habit which we justify with religious ideas or with whatever crumbs of public spending may fall from our masters' tables. At root this deep wiring of subservience is an inheritance that does nothing to serve the interests of ordinary people. It predisposes us to patterns of autocracy, taxation and exploitation of the many by the few. It's a pathology that makes us easily managed.

Could this slave-wiring be the reason why we human beings are often so willing to give our power away to others or to fawn in adulation over royals, idols and superstars? Is it why we ogle at the lives of the rich and famous on 1 percent TV shows that help us to keep up with the Kardashians and the real housewives of wherever next? Is this deep wiring the reason why we allow ourselves to be hijacked by elites, whether through corrupted systems of money and banking, unjust patterns of employment, forcible enslavement, police brutality or political tyranny? The claim of the Sumerian and Babylonian accounts of our creation is that that is precisely what we were engineered to do.

When we human beings learn to disassemble our subservience towards elites and shake off the structures by which we enslave one another; when we learn to labor for the common good instead of for higher powers, then we will have upgraded ourselves to a higher consciousness and a better way of being.

In the Gospel Jesus aims his fire directly at that slave programming when he says:

"You know how among the gentiles (i.e. the peoples of the world) *their leaders lord it over them, and how the great men make their authority felt? It must not be so among you..." (Matthew 20:25)*

Or again, this time with reference to the Jewish religious authorities of his day, Jesus says:

"They tie up heavy burdens and lay them on people's shoulders, but will they lift a finger to help them? No! Everything they do is done to attract attention...You have only one Master and you are all brothers [and sisters]."

If the Sky People regarded human beings as a slave species to serve them, Jesus' attitude is something quite different. At the beginning of his ministry Jesus announces that he has come to give power to the afflicted and oppressed and to set captives free. *(Luke 4:18)* He refers explicitly to release from slavery when he says: *"Then you will know the truth and the truth will set you free...And if the son sets you free then you are free indeed!" (John 8:32,36)*

In a similar vein the writers of the New Testament affirm that our Heavenly Father sees human beings not as slaves but as beloved, dearly beloved, little children. To his deputies and followers Jesus says, *"I have called you 'friends'. Servants don't understand what their masters are about. But I have called you friends."*

When we bring our slave-programming into our relationship with God it pushes us into patterns of seeking to please God through sacrifices of prayer, faith, obedience and service, in order to obtain his favor or assistance. We habitually forget that Jesus has completely reframed what *"serving God"* really means. In the parable of the sheep and goats Jesus teaches that we serve God not by slaving for a divine superior but by serving and caring for one another, and in particular serving those weaker than ourselves. It is a totally different paradigm. All the New Testament writers echo and re-express that thought.

In short, the old programming of Sumerian slaves bringing sacrifices to a Sky People, or of the children of Israel slaving to the demands of a morally inscrutable divine master, is replaced in the New Testament by a vision in which loving, compassionate humanity expresses and participates in the true nature of God.

According to Genesis, the Sumerian tablets and the Popol Vuh, not only were we mentally programmed for a life of slavery but our mental capacities and powers of perception were downgraded. To maximize our usefulness to our overlords, our brains' default settings were deliberately turned down.

These mythological explanations may be mind-boggling, yet they reflect contemporary questions about human behavior and human brains. We only have to look at our brains to ask, *"Why is 90 percent of our brain not used? Is there more within our brains waiting to be switched on?"*

CHAPTER THIRTEEN

PLEASE SIR, I WANT SOME MORE!

Some light bulb moments are more dramatic than others.

In 1980 Orlando Serrell was hit on the side of the head with a baseball. He was knocked out by the blow but, being an eager sportsman of ten years old, he didn't let his brief blackout prevent him from finishing the game once he had come around. Orlando suffered headaches for a year after the accident. When the symptoms receded, Orlando realized that something had changed. Today if you put any date to him, any date in any year, Orlando can tell you the day of the week you're referring to. And he does this without fail. Ask him how many times the 15th April has fallen on a Wednesday and he will just know. Pluck out two random dates from different years and he will instantly tell you how many days elapsed between the two dates. Ask him what happened on February 11th 1983 and he will tell you, *"It was a Friday. It rained and I got a pepperoni sausage pizza from Dominos."*

This incredible capability was switched on by a blow to his head in 1980 and has remained on ever since. Orlando exhibits what neurological scientists call Acquired Savant Syndrome.

The pattern of Acquired Savant Syndrome is that a person who previously had exhibited no extraordinary mental or creative skill suddenly finds a higher capacity has been released by a central nervous system (CNS) injury.

Ben McMahon grew up down the road from me in Victoria. At some point during his education he did a couple of years' study of Mandarin, but it didn't take. He was never able to speak it convincingly or fluently. That is until 2012 when he nearly died in a car crash. Ben suffered a head injury and was in a coma for a week.

When Ben finally opened his eyes, he beckoned the nurse

and said, *"Excuse me, nurse, I feel really sore."* He said it in fluent Mandarin. Confused by the sounds that had come out of his mouth he indicated that he wanted a pen and paper. He wrote down, *"I love my mum. I love my dad. I will recover."* The note was written in Mandarin.

Thankfully Ben's ability to speak in his native English returned after three days. Yet his Mandarin skills persisted and have remained ever since. And it wasn't the half-learned schoolboy Mandarin that he had before. Ben was fluent. In fact he was so fluent that he soon moved to Shanghai, where for a time he hosted a TV game show and then went on to study at a Chinese University.

Between 1996 and 2000 Dr. Bruce Miller, Professor of Neurology at the University of California, catalogued twelve cases of Acquired Savant Syndrome. Strangely these were cases where musical and artistic skills were released or dramatically enhanced by the onset of frontotemporal dementia. His research team hypothesized that selective degeneration of a particular part of the brain had resulted in *"decreased inhibition of visual systems."* Now that's an interesting phrase. In 2005 a research team led by Professor Dr. Mark Lythgoe of University College London demonstrated that degeneration in a particular area of the brain *"may release untapped cognitive abilities."*

In an amazing bit of research in 2006, a team led by Dr. Drago set up an experiment to judge the elevated artistic skills of an artist whose painting style had undergone a metamorphosis during the onset of frontotemporal degeneration.

Without telling them anything about the artist or her story, the researchers invited a number of art judges to assess the changes in the artist's work. The judges were shown eighteen pictures from before the artist had shown any symptoms of degeneration, six paintings from the time that her symptoms were just emerging, and sixteen paintings from the time that she was fully symptomatic.

Without knowing the dates of the paintings or anything about the artist's diagnosis the art judges critiqued each painting. Their findings gave the same report, as the degeneration progressed, her skill level was lifting in a way that reflected in enhanced artistic technique. The growth in skills was described by the research team as evidencing a *"disinhibition"* of part of the brain.

There's that language again. Apparently, entirely by accident, an inhibitor was being switched off. Hearing that we have to ask, *"What in the world is an inhibitor doing in our brains?"* Mind bending though it is, Genesis, the Mesopotamian tablets and Popol Vuh all propose the same answer. It was put there.

Writing in 2018, Dr. Darold Treffert, of Marian University, a research director in psychiatry, asks the obvious question: *"Is it possible that such dormant potential resides in all of us?...The challenge of course, if that is so, is how to tap those hidden abilities without having endured some CNS catastrophe."*

Once you have seen and heard a few cases of Acquired Savant Syndrome you cannot just go back to business as usual. A light has been switched on. Contemporary neurological science and the voices of our ancient mythologies compel us to ask how we can *"disinhibit"* our human capacities. If we see this potential getting unlocked by accident through a CNS injury, then we surely have to explore how such potential may be tapped deliberately and move our default settings a little higher.

In the Gospels and New Testament Jesus reveals a God who loves human beings as we are but does not wish us to remain as we are. All Jesus' teachings are about elevating the human condition. His teachings continually invite us to be far more than we have yet imagined ourselves to be.

The New Testament letter of *II Peter* calls on its hearers to learn how to *"participate in the Divine nature and escape the corruption in the world."* The exhortation that follows is to goodness, understanding, self-control, perseverance, devotion, kindness and love. *(II Peter 1)* These are given as means by which we can elevate ourselves to a higher way of being.

In a similar vein the Apostle Paul's writing teaches us to transcend our biological wiring and be led by the Spirit of God, resident within us. When we allow this spiritual aspect to drive us, the change in our being will be manifested in *"love, joy, peace, patience, kindness, goodness, faithfulness, gentleness and self-control." (Galatians 5:22)*

In his most famous teachings – The Sermon on the Mount and the Sermon on the Plain, Jesus teaches ways of being that are not only healthy at a societal level but which avoid behaviors that have the potential to ruin any life. Jesus referred to that ruination Gehenna – and it was the refuse tip outside the city of Jerusalem.

The destructive behaviors from which Jesus calls us to disengage include writing people off, adultery, being led by lust, infidelity to our spouses, dishonesty, revenge, selfishness, meanness, hatred and abuse of children. In each instance Jesus appeals to a higher faculty, one resident in every human being, our ability to foresee consequences.

Furthermore, the Gospels show us a Jesus who lives unafraid of the powers of that time. He brings the same freedom and self-esteem to many who were rejected and down-trodden. He sets worshippers of God free from their attachment to priesthoods and elites. He effects healing for the sick and those troubled by evil spirits. He brings the dead to life and gives joy to the mourning.

A little more weirdly, Jesus sources money from a fish, generates a huge feed of fish from an empty basket, and a huge catch of fish from empty water. He walks on the same water, teaches his friend to do the same, and arrests a violent storm simply by speaking to it.

So when Jesus pauses from all that activity and declares, *"Anyone who believes in me will do the things I have been doing – these things and even greater things – because I am returning to the*

Father..."

...what are we to make of that. What *"greater things"* could there possibly be?

This incredible saying invites us to imagine a way of being where all kinds of *"inhibitors"* have been switched off. Jesus lived freely and authentically in the face of intimidation, conflict and hostile powers. How might that look for us? Jesus' miracles overturned all our conventional understandings of what is and is not possible. How might that look for us? If he can bring healing to bodies and minds and elevate those around him, how much more of that might we do? He did all this through an intimate, conscious rapport with his Heavenly Father. What if we were to enjoy the same experience? If Jesus represents what a human life might look like with the slave programming and all the inhibitors switched off, then I want some more of that!

And what about our *"untapped cognitive abilities"* – to use Dr. Mark Lythgoe's phrase? What might a human life look like with our brains on a higher setting?

Popol Vuh speaks of an earlier setting whereby we can see beyond what is local and physical. What would life be like with powers of perception switched on at that level? The book of Genesis and the cuneiform tablets speak of a previous setting engineered for fantastically long life-spans. With what confidence, patience and courage might we live if we were equipped with super-robust health and longevity? Before Babel, Genesis portrays a world of mutual understanding and easy communication. How good would a twenty-first century be if it were built on concord like that?

When Jesus said, *"These and even greater things will you do..."* I cannot think of a greater invitation than that to explore – and not for the sake of curiosity, but to transform our lives.

CHAPTER FOURTEEN

CONCLUSION – WHO AM I?

We bought the shipping crate as our accommodation for guests. But it had come in handy as a quiet spot for study and my private place of prayer.

I began in Genesis, studying up for my next sermon at the Church on the Range. But the anomalies had taken me on a journey around the world, away from the hills and valleys of Victoria, on a tour of ancient Sumeria, Babylonia, Greece and Egypt; from India to South Africa, Peru, Bolivia and Mexico.

I felt like Neo waking up from his illusory life in the Matrix. Now I understood that what I had previously regarded as glitches in my Christian worldview were actually flashing lights signaling an altogether different meta-narrative – the narrative to which these texts actually belong. When I began my hermeneutical exercise I could never have imagined where a plural Elohim was – or should I say were – going to lead me!

It was a pleasure to be back in Victoria, in the quiet, shady spot at the end of our driveway, enjoying the warmth of my shipping crate seclusion, readying it for guests and sorting through a floor full of notes. Here, just for a few more days, I could sit knee to knee with Jesus and ask him to help me make sense of all I had been learning.

In the Gospel of John, Jesus promises that the Spirit will remind us, whenever needed, of the words and truths of Jesus. This is the confidence every preacher banks on as they carry the questions of their communities and minister to the needs of others. It's what they do. So in my seclusion I was in a familiar space with Jesus and my questions for company.

I didn't find it hard to believe that our galaxy might be more densely populated than we have generally been taught – and that it may have been seeded with and by people who look similar to us. But where was Jesus in this picture – and what did he think about our alien brothers and sisters? Did he agree with Fr. Funes and Fr. Consolmagno at the Vatican Observatories?

In my mind I could hear Jesus say, *"I have others who are not of this fold. I must bring them also."*

What about these others then? A *"brother alien"* is one thing but would God really have allowed ET species to trespass on the soil of our beautiful blue-green planet, plunder its resources, meddle in the life of its flora and fauna, genetically modify us and exploit us as we do with livestock? Surely our God wouldn't permit interstellar visitors to rule over the communities of our prehistoric ancestors as *"heavenly kings"* or *"gods"?* OK, so God allowed us to do it to each other as generations of colonizers and colonized, but would God really have allowed a non-terrestrial species to do that to our distant ancestors? Surely God wouldn't leave us at the mercy of alien marauders and interlopers like that? If God really loved human beings, wouldn't he step in to save us from such false gods?

Jesus said, *"All who came before me were thieves and robbers…The thief comes only to plunder, and kill, and destroy. I have come that [my sheep] may have life and have it in all its fullness." (John 10.8a,10)*

I love what I see in Jesus. However, with a Bible scribed by ancient authors who were gradually feeling their way from foreign worldviews to the monotheism of Sunday school religion, how could I recognize God's authentic revelation? What could Jesus show me within the pages that was pure, unalloyed and clear in depicting the Father as he really is?

Jesus said, *"Don't you know me…even after I have been among you such a long time? How can you ask show me the Father? Anyone who has seen me has seen the Father." (John 14.9a)*

So where in the Bible is that vision of the Father most developed? Where is it the clearest and least spun by the writers' perspectives at the time?

The writer to the Hebrews says, *"In the past God spoke to our ancestors at many times and in diverse ways through the prophets. But now..."*

That *"but now"* means something new and better has come; something that contrasts with all that went before and is of a different order of magnitude...

"...But now in these last days he has spoken to us through his Son, whom he appointed to inherit everything and through whom he made the universe. The Son is the radiance of his glory, the exact representation of his being." (Hebrews 1:1–3a)

Now I said, *"And who am I? How can I be your creation, your Father's son, if the truth is that some other flesh and blood species had a hand in engineering me?"*

Jesus said, *"Flesh gives birth to flesh. Spirit gives birth to spirit." (John 3:6)*

Wow! The Gospel of John really seemed up for my questions! It remained calm and unruffled as I continued my interrogation.

"So what are we doing here with 90 percent of our brains switched off? What might we be capable of if we could learn to change the default settings without resorting to car crashes and comas? Can we switch our brains on? Surely, Jesus, you don't want us scratching around with neural slavery settings still determining what we imagine ourselves to be capable of?"

Jesus said, *"Anyone who believes me will do what I have been doing. These and even greater things will he do..." (John 14:12)*

I still wondered what Jesus knew when he was on Earth. Did he know the things I was now discovering? Or was he totally immersed in our humanity, beginning as a baby, totally dependent on his mother and father to nurture and teach him? Was this immersion so entire that he received his insights from the Holy Spirit on a need-to-know basis? Or perhaps Jesus knew more than he ever spoke and shared of what he knew, on our need-to-know basis?

The words of Jesus came to mind, which said, *"I have much*

more to say to you, more than you can now bear. But when he, The Spirit of truth comes, he will guide you into all truth...He will take from what is mine and make it known to you." (John 16: 12, 13a, 15b)

I cast my eye over my notes as I begin to pick up the months' worth of piles of paper from the floor of the cabin and reflect on the implications of my mythological travels.

Disentangling God from the activity of others has shown me a better God than I knew before. I want to know that true Divinity more fully and more authentically – and without viewing him through the lens of fear and servility programmed into us by others and by centuries of religion.

My universe has grown. It is more mystifying and more populous than I thought before. I have a bigger family than ever I knew and can look forward to all that my wider family's technology may have to offer. How good will it be to fuel the world with zero-point energy and set our planet free from our dangerous and expensive enslavement to nuclear energy and fossil fuels.

And if our galactic neighbors can get here, presumably there are subspace technologies that might transform our own relationship with the universe. That's an exciting future. It all brings me back to the question of the Psalmist as he looked out at the stars of heaven and asked, *"What is man that you are mindful of him; the son of man that you care for him?"*

What are human beings? And what are we capable of? I want to know how our lives can be with better tapped brains and old inhibitors switched off. I want to really unpack all that Jesus gave us to unravel our subservient and slave programming. I want to know what a free, sibling society, and love for all our brothers and sisters might look and feel like.

To acknowledge, even for a moment, that you and I live on a beautiful blue-green dot that can be hit, blasted and rebooted by a random comet or solar flare, just like that, makes me want to live in tune with the realm of eternity, living my life here to the very fullest, unafraid and ready for the next step.

After months of torsion and traction I am very happy to be walking stronger – and not living beholden to the discipline of having every *i* dotted and every *t* crossed by next Sunday morning. With the joy of that freedom my appetite has come alive.

I pick up my Bible, which is open at John's Gospel, chapter ten. *"I have come that [my sheep] may have life and have it in all its fullness."*

If God is as he appears in Jesus then there is joy, love, power and freedom itching to be taken hold of and expressed in a world of people made fully alive.

I stand up and test my leg. It's feeling good and strong. Better than before even! I may even be ready for another round of Ultimate Frisbee! More importantly, I think I'm ready for the next part of the journey. But right now, this minute, my family is at the door, the car is ready to leave and my editor is texting me. It's time for that press conference!

EXPLORING THE WORLD OF HIDDEN KNOWLEDGE

Axis Mundi Books provide the most revealing and coherent explorations and investigations of the world of hidden or forbidden knowledge. Take a fascinating journey into the realm of Esoteric Mysteries, High Magic (non-pagan), Mysticism, Mystical Worlds, Goddess, Angels, Aliens, Archetypes, Cosmology, Alchemy, Gnosticism, Theosophy, Kabbalah, Secret Societies and Religions, Symbolism, Quantum Theory, Conspiracy Theories, Apocalyptic Mythology, Unexplained Phenomena, Holy Grail and Alternative Views of Mainstream Religion.

If you have enjoyed this book, why not tell other readers by posting a review on your preferred book site? Recent bestsellers from Axis Mundi Books are:

On Dragonfly Wings

A Skeptic's Journey to Mediumship

Daniela I. Norris

Daniela Norris, former diplomat and atheist, discovers communication with the other side following the sudden death of her younger brother.

Paperback: 978-1-78279-512-4 ebook: 978-1-78279-511-7

Inner Light

The Self-Realization via the Western Esoteric Tradition

P.T. Mistlberger

A comprehensive course in spiritual development using the powerful teachings of the Western esoteric tradition.

Paperback: 978-1-84694-610-3 ebook: 978-1-78279-625-1

The Seeker's Guide to Harry Potter

Dr Geo Trevarthen

An in-depth analysis of the mythological symbols and themes encountered in the Harry Potter series, revealing layers of meaning beneath the surface of J K Rowling's stories.

Paperback: 978-1-84694-093-4 ebook: 978-1-84694-649-3

The 7 Mysteries

Your Journey from Matter to Spirit

Grahame Martin

By simply reading this book you embark on a journey of transformation from the world of matter into spirit.

Paperback: 978-1-84694-364-5

Angel Healing & Alchemy

How To Begin Melchisadec, Sacred Seven & the Violet Ray

Angela McGerr

Angelic Healing for physical and spiritual harmony.

Paperback: 978-1-78279-742-5 ebook: 978-1-78279-337-3

Colin Wilson's 'Occult Trilogy'

A Guide for Students

Colin Stanley

An essential guide to Colin Wilson's major writings on the occult.

Paperback: 978-1-84694-706-3 ebook: 978-1-84694-679-0

The Heart of the Hereafter

Love Stories from the End of Life

Marcia Brennan

This book can change not only how we view the end of life, but how we view life itself and the many types of love we experience.

Paperback: 978-1-78279-528-5 ebook: 978-1-78279-527-8

Kabbalah Made Easy

Maggy Whitehouse

A down to earth, no-red-strings-attached look at the mystical tradition made famous by the Kabbalah Center.

Paperback: 978-1-84694-544-1 ebook: 978-1-84694-890-9

The Whole Elephant Revealed

Insights Into the Existence and Operation of Universal Laws and the Golden Ratio

Marja de Vries

An exploration of the universal laws which make up the dynamic harmony and balance of the universe.

Paperback: 978-1-78099-042-2 ebook: 978-1-78099-043-9